T0134336

Rainer Grießhammer | Bettina Brohmann

How Transformations and
Social Innovations Can Succeed

Transformation Strategies and Models of Change
for Transition to a Sustainable Society

In cooperation with:
Dierk Bauknecht | Martin David (Institute for Advanced Study in
Humanities Essen) | Dirk Arne Heyen | Inga Hilbert | Lucia Reisch
(Copenhagen Business School)

 Nomos

Translation: Christopher Hay

The Deutsche Nationalbibliothek lists this publication in the
Deutsche Nationalbibliografie; detailed bibliographic data
are available on the Internet at http://dnb.d-nb.de

ISBN   978-3-8487-2741-4 (Print)
        978-3-8452-7085-2 (ePDF)

British Library Cataloguing-in-Publication Data
A catalogue record for this book is available from the British Library.

ISBN   978-3-8487-2741-4 (Print)
        978-3-8452-7085-2 (ePDF)

Library of Congress Cataloging-in-Publication Data
Grießhammer, Rainer / Brohmann, Bettina
How Transformations and Social Innovations Can Succeed
Transformation Strategies and Models of Change for Transition to a Sustainable
Society
Rainer Grießhammer / Bettina Brohmann
74 p.
Includes bibliographic references.

ISBN   978-3-8487-2741-4 (Print)
        978-3-8452-7085-2 (ePDF)

1. Edition 2015
© Nomos Verlagsgesellschaft, Baden-Baden, Germany 2015. Printed and bound in Germany.

Preface

Germany's energy transition (*Energiewende*), as a targeted intentional transformation, is a topic of ongoing debate. Easy, then, to overlook the general lack of progress on another transformation, namely sustainable development, proclaimed by the international community at the Rio Summit back in 1992.

Concepts of transformation or transition offer great promise for more sustainable policy development, especially since it has become apparent in recent decades that one-off technical solutions such as better energy efficiency, end-of-pipe technologies and classic policy instruments – despite having successfully mitigated many environmental and sustainability problems – rarely result in structural change.

Transformation concepts are also of interest in relation to systemic innovations, for they explain why some initiatives – such as carsharing – have been successful while others, notably the Transrapid high-speed rail system, have spectacularly failed. These concepts show why, in future, emphasis needs to be placed not only on technical but also on social and institutional innovations.

In recent years, there has been a plethora of publications on transformations and transition management. This book provides a systematic evaluation of these works and considers their practical relevance, drawing on numerous examples from the energy transition, stakeholder cooperation (for instance, the energy transition committees and chemical policy initiatives) and the role of the municipalities (such as Green City Freiburg)[1].

Additionally to science, the book is therefore intended for actively engaged individuals in politics, administration, business, civil society and

---

1 The recommendations and practical information presented in this book, are based on the results of the UFOPLAN project „Transformationsstrategien und Models of Change für nachhaltigen gesellschaftlichen Wandel". We kindly thank Dierk Bauknecht, Dirk Heyen, Inga Hilbert and Andrea Droste, our project partners Lucia A. Reisch, Claus Leggewie and Martin David as also the supervisor of the Federal Environment Agency Alexander Schülke for their inputs, critical guidance and discussion.

local networks who are key decision-makers, multipliers or communicators on environmental and sustainability issues.

Various questions of a practical nature arise in this context. Who initiates and shapes a transformation, and at which levels can it be driven forward? How can the right stakeholders be identified? How should we manage uncertainties and the exploratory processes that will doubtless be required? The book provides guidance here. Transformations cannot be planned in detail, but it is possible to define the major goals and adopt the directional decisions that become necessary in practice. In doing so, transition stakeholders striving for success should be guided by Wilhelm Raabe's old adage – "Look to the stars and keep an eye on the alleyways!"

Table of Contents

1. Introduction

Those who are willing find ways. Those who are unwilling find excuses.
Albert Camus

Production and consumption patterns in Germany and many other developed economies are resource-intensive and harmful to the environment, and despite all the initiatives for a new social and ecological paradigm, they remain unsustainable. The adoption of a Western lifestyle by nine billion people in every nation of the world would far exceed the Earth's ecological carrying capacity (Rockström et al. 2009, Steffen et al. 2015). Global change is worsening, leaving little time to establish globally responsible lifestyles and economic patterns which would be sustainable in the long term. For instance, developed economies such as Germany must drastically reduce their greenhouse gas emissions (by 80-95% by 2050) in order to at least buffer catastrophic climate change (IPCC 2007). The consumption of natural resources by developed economies should also be reduced by approximately 80% (Weizsäcker et al. 2009). Stand-alone technical solutions, such as increases in energy efficiency (e.g. domestic appliances), and conventional (environmental) policy instruments (e.g. Germany's Waste Avoidance and Waste Management Act – *Abfallgesetz*) have yielded specific progress in recent decades and will continue to be necessary. Also in the field of greener technologies progress was archived. Typical strategies have been end-of-pipe technologies (e.g. catalytic converters for cars; flue gas desulphurisation (FGD) technologies in coal-fired power plants) or drop-in solutions in the chemical industry (e.g. the replacement of ozone-depleting CFCs with other chemicals). However, ambitious sustainability targets cannot be achieved by these developments alone, because the systems themselves (the energy supply, housing, transport, agriculture and food production systems) – which are resource-intensive and leave a large ecological footprint – and generally high levels of per capita consumption have remained essentially unchanged. This is reaffirmed at the national and international level by the calls for ambitious Sustainable Development Goals (SGDs) and a green economy.

In Germany and other developed economies, draft strategies for a radical transformation and *Wenden* – "transitions" – have long existed for individual sectors (see the following publications: Ende oder Wende [politics] (Eppler 1975); Energiewende [energy] (Krause et al. 1980); Landbau-Wende [agriculture] (Bechmann 1987); Chemiewende [chemicals] (Grießhammer 1992); Verkehrswende [transport] (Hesse 1995); Ernährungswende [food] (Eberle et al. 2006)), mainly emanating from the environmental movement. The fall of the Berlin Wall and German reunification, too, are often known collectively as the *Wende* and 1989 as the *Wendejahr* – the year which marked a turning point for Germany and the start of sweeping political change at international level.

What might a future societal transformation process – also described as the "quest for a new social paradigm" – look like, and how can it be managed? These questions still remain unanswered. Nonetheless, it is a challenge which must be addressed by (environmental) policy, which is seeking to identify how incentives (governance) can be provided for successful societal change and what might contribute to creating more sustainability and a culture of sustainability.

Governance of societal change requires, on the one hand, a system perspective, which aims to transform entire sectors (e.g. the energy supply). On the other, it also requires an analytical approach which distinguishes between relevant fields of action and subsystems in order to identify entry points for initiating change and interactions between the subsystems.

Experiences are provided by historical transformations, for example the first and second Industrial Revolutions. The first Industrial Revolution was driven by the invention of the steam engine, the spinning jenny and the power loom. The second was powered by electricity, oil (a flexible energy source), cars and mass production. In addition to technological innovations, the Industrial Revolutions resulted in changes and upheavals in many sectors of society: working conditions, production processes, property ownership, and social and political relations. Based on these experiences, recommendations on the active governance or, at least, the influencing of desirable, "intentional" transformations are discussed. Transformations, in this context, are defined as deep-seated and long-term cultural, social, technological, economic, infrastructural, production- and consumption-related co-evolutionary changes affecting society as a whole across sectors and subsystems, not simply one-off technical or social policy solutions. The literature describes various concepts and proposals, ranging from the "Great Transformation" (WBGU 2011) to networked transformation or transitions in individual sectors. One of the research community's aspirations, in this context, is to make key recommendations for the shaping of intentional transformations and for the fostering of societal and socio-technical innovations which go far beyond the strategies and practices hitherto pursued in environmental and sustainable development policy. This instantly piques our curiosity in two respects: on the one hand, scholarly endeavour in the field of environmental and sustainability policy in recent decades has hardly been noted for its great practical relevance; on the other, the environmental and sustainability policy pursued in recent decades has only achieved limited success – reason enough to look at the issue more closely. In short, the effort pays off.

The basic concept of "transformations/transitions" is summarised in Chapter 2. It is supplemented by information on the complex multi-level processes in politics; on key aspects of cultural change; on the particular significance of time as a factor in transformations; and on tipping points that may potentially be triggered in social systems (Chapter 3). The findings were published in diverse reports (Bauknecht 2015; Reisch & Bietz 2014; Brohmann & David 2015) and working papers (Heyen 2013; David & Leggewie 2015). These outcomes formed the basis for the present report. They were supplemented by parallel case studies on the role of municipalities, with particular reference to Green City Freiburg (Grießhammer & Hilbert 2015), and the effectiveness of stakeholder cooperation, based on the example of the 400 energy transition committees (*Energiewende-Komitees*) set up in Germany in 1986 immediately after the Chernobyl disaster (Grießhammer et al. 2015; Brohmann 1996).

The project's key findings on innovation management and the practical governance of transformations were summarized, and are illustrated with examples (Chapter 4). Based on the previous considerations, policy recommendations and the need for further research are summarised in Chapter 5. As a conclusion, governance challenges for running and future transformations are summarized (Chapter 6).

This book is intended for actively engaged individuals in politics, administration, business, civil society and local networks who are key decision-makers, multipliers or communicators on environmental and sustainability issues.

2. Understanding transformation

Transformations lead to structural, paradigmatic changes in society, affecting culture, values, technologies, production, consumption, infrastructures and politics. The processes are co-evolutionary, concurrent or consecutive and take place in various areas or sectors, possibly influencing, reinforcing or weakening each other, sometimes to a considerable extent. A key characteristic of a transformation is that the processes intensify over time and cause fundamental and irreversible changes in the regime; this is known as a paradigm shift. Transformations may be unplanned or intentional. They may take several decades and progress at varying speeds.

In contrast to the historical transformations (notably the first and second Industrial Revolutions) which were not planned in a targeted manner, it is now assumed that intentional transformations (e.g. Germany's energy transition) can, to a large extent, be influenced to ensure that they move in the desired direction and can be accelerated, even if they cannot be managed in detail. This assumption is based on the wealth of experience and information now available about complex steering, governance and strategic approaches. However, intentional transformations also create specific challenges, such as achieving a social consensus on the aims of the transformation, the intended acceleration of the transformation, the action that should be taken in relation to potentially open-ended technological and social innovations, and measures to overcome resistance to the transformation. Among other things, resistance is characterised by infrastructural and technical path dependencies, fears of change, vested interests, the predominant production and consumption culture, an overemphasis on growth, and short-term thinking.

Transformations can be differentiated according to their size: major (e.g. first and second Industrial Revolutions and the Great Transformation proclaimed by WBGU, WBGU 2011), intermediate (e.g. the German energy transition), and smaller-scale (e.g. digital publishing and reading); there are also systemic innovations without radical structural change (e.g. the evolution of cycling in recent decades; see Section 3.3.1).

Major and intermediate transformations can consist of several or many smaller-scale transitions. In Germany's energy transition, for example, a distinction can be made between the transformation of the electricity supply, the building stock, the transport system, and many smaller transformations in the field of energy efficiency. These parallel transformations can influence each other, either positively or negatively. Even if the goal is the same, conflicts or undesired impacts can occur, e.g. due to multiple access to limited resources. Examples of reciprocal interdependences are the rise in photovoltaic and wind power and the effects on the electricity grid and demand for storage capacity. Biomass is an example for a scarce resource for which there is high demand in a number of sectors, notably buildings/heating, transport, food and agriculture.

The terms are not clearly defined, and they may be construed in different ways. For example, one point of contention is whether, due to the massive changes brought about since the 1980s by information technologies and biosciences, the world is already undergoing a third Industrial Revolution. And indeed, it is much easier to describe a transformation after the event than to determine, at the time, whether a society is in the throes of transition, when it began, and how significant it is likely to be. In historical and, indeed, present-day transformations alike, society and the actors themselves were, or are, often unaware of their own participation in a transformation.

A transformation can take place, perhaps incrementally, over many decades, or it may have an abrupt and dramatic trigger (e.g. a war or a large-scale natural disaster such as a major volcanic eruption). A transformation may unfold in an unplanned manner (e.g. anthropogenic climate change) or be intentional. One example of an intentional transformation is Germany's energy transition, also known as the energy turnaround (*Energiewende*), which – if it is successful – has the potential to be the germ cell of global energy system transformation.

Transformations may move in desired directions (e.g. the energy transition) or in undesired directions, and may be associated with social injustice and massive environmental pollution (e.g. the first and second Industrial Revolutions). This book focuses on intentional transformations to sustainability. However, here too, it is important to analyse transformation processes which move society in an unwelcome and unsustainable direction, such as the feared consequences of the adoption of the Transatlantic Trade and Investment Partnership (TTIP).

The concept of sustainable development, adopted by the international community at the Rio Summit in 1992, displays key elements and drivers of an intentional major transformation. The Rio Summit was preceded by decades of international debate about global environmental policy (e.g. the Stockholm Conference in 1972), poverty reduction, equitable resource governance, debt relief/debt restructuring strategies for developing countries, and justice in world trade. Key international agreements were adopted, notably the United Nations Framework Convention on Climate Change (1992)[2] and the Convention on Biological Diversity (1992)[3]. However, as the lengthy and difficult decision-making processes on further climate change mitigation and on a commitment to more comprehensive Sustainable Development Goals (SDGs) show, the required structural changes in the prevailing production and consumption system are contentious.

A basic taxonomy of the evolution and progression of transformations is provided by the **multi-level perspective** of transition management. Developed on behalf of the Government of the Netherlands in 2000 (Loorbach 2007), transition management operationalises a range of information about social transformation processes and their governance and translates it into a practical policy instrument. Various methods (e.g. scenario analysis, strategic niche management) can be combined within the toolkit.

The term "transition management" hints at this practicality, but should not be construed as meaning that the use of this methodology genuinely allows transformation processes to be managed in the narrower sense. The multi-level perspective is fundamental to an understanding of transition management (Figure 1). In this perspective, there are three levels – niches, regime and landscape[4] – within a system, with interactions between them.

---

2 United Nations Framework Convention on Climate Change, New York 1992
   https://unfccc.int/resource/docs/convkp/conveng.pdf
3 Convention on Biological Diversity, Rio de Janeiro 1992
   https://www.cbd.int/doc/legal/cbd-en.pdf
4 The key terms landscape, regime and niche are generally translated as Landschaft, Regime and Nische in the German literature. From the authors' point of view, these literal translations are somewhat misleading and cause unnecessary confusion. The term Landschaft is incomprehensible without further explanation, so the authors prefer to use the term Globale Lage (global level). In German, the word Regime tends to have undertones of authoritarianism; the preferred translation is therefore Vorherrschendes System (prevailing system). For the English version the original terms are kept by.

Figure 1   The multi-level perspective and its three tiers

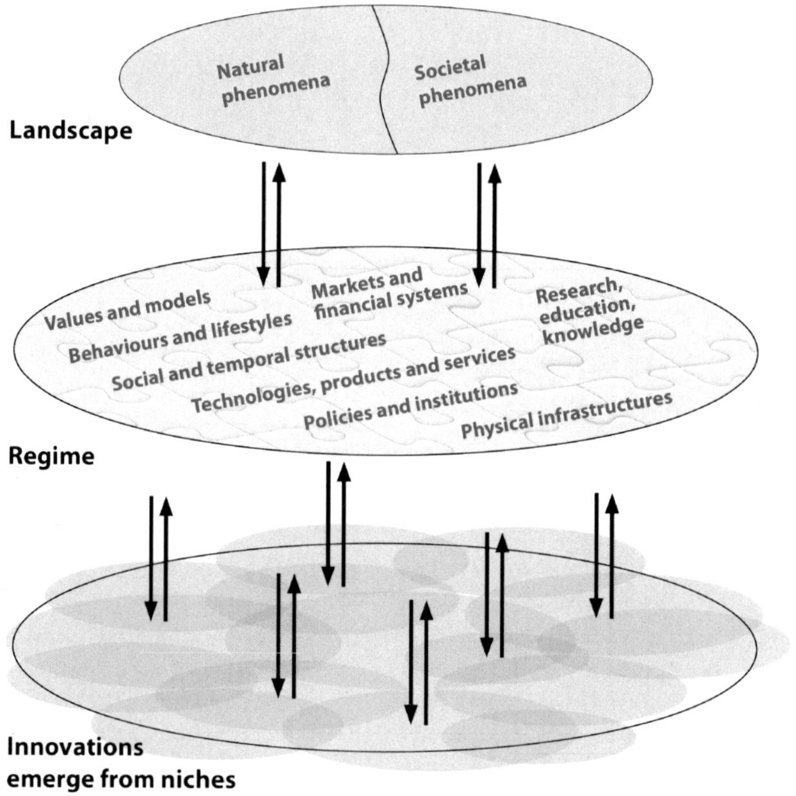

Source: Own depiction, adapted from Geels 2002

The regime is located at the meso level and is the (pre)dominant model for problem-solving by society and economy (e.g. in relation to electricity supply). It comprises a network of institutions and actors with established solutions (e.g. large conventional power plants, base load power plants, nuclear plants and lignite, major energy suppliers, overcapacity, etc.).

Developments at the landscape e.g. climate change, Fukushima, oil price rises, artificial scarcity and political conflicts, can exert pressure for change on the regime and thus help to ensure that innovations emerge from the niche (i.e. the micro level) and become the new core of the regime (e.g. photovoltaics, wind power). Niches are "protected spaces" in which technological, market, social or regulatory innovations emerge, with high potential to effect change within the regime. One example for niche developments in protected spaces is Car sharing, which first evolved as local initiatives run by non-profit associations and staffed by unpaid volunteers (members).

Niche actors (also known as pioneers of change) and targeted innovation management aimed at encouraging the development of these niches (strategic niche management, Smith & Raven 2012) play a key role in transition management. This also applies to the anticipation of fundamental changes in the landscape and regime via scenario processes and to the participatory development of visions for the future system.

Based on innovation research, the **progression** of transformations **over time** can typically be described as an S-curve with the following four phases:

1. Predevelopment, with numerous innovations (mainly in niches)
2. Take-off, with initial changes
3. Breakthrough, with structural transformation resulting from the accumulation and intensification of changes
4. Stabilisation, in which a new dynamic equilibrium and a new regime emerge.

Figure 2 Typical progression of transformations

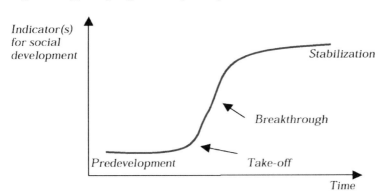

Source: Kemp & Loorbach 2006

One-off major events, such as Fukushima, may bring matters to a head. However, the German Government's policy decision in favour of **energy system transformation** and an accelerated nuclear phase-out was only partly the outcome of record levels of public support in Germany for a

17

rapid and complete nuclear phase-out after the Fukushima disaster. A more significant factor was probably the existence of a well-established and powerful anti-nuclear movement in Germany and the fact that the expansion of renewables had been well-prepared – politically, structurally and technically – for years, such that it was a genuine and politically viable alternative to "business as usual". Key contributory factors prior to Fukushima were the oil crisis (1974), the Oeko-Institut's publication of an initial strategy for the energy transition (Krause et al. 1980), anti-nuclear protests, the onset of climate change, Chernobyl (1986), Germany's 100,000 Roofs Programme[5] (1999) and the German Renewable Energy Sources Act[6] (2000).

---

5 Guideline to promote Photovoltaic Power Plants (300 MW) by „100.000-Dächer-Solarstrom-Programm" January 15th 1999 (BAnz. S. 770)
   https://de.wikipedia.org/wiki/100.000-D%C3%A4cher-Programm
6 German Renewable Energy Sources Act (EEG). Bundesgesetzblatt Jahrgang 2000 Teil I Nr. 13, ausgegeben zu Bonn am 31. März 2000, S. 305-308.

3. Managing innovation and transformation

In the transformation literature, the fundamental challenges, structural impediments, basic success factors, and the need to link developments in various subsystems are well-described. However, key aspects are often neglected, such as the analysis of ongoing transformations, economic dimensions, the important role played by businesses, the issue of growth, funding, conflict management, political and social interests/interest groups, and international cooperation.

Various attempts have been made to develop and formulate political and practical solutions, but the methods and tools proposed tend to lack coherence. This is hardly surprising, given the complexity and the lack of historical examples of well-managed transformations. The theoretical foundation for understanding systematic interactions among subsystems is slim; moreover, it is doubtful whether understandings gained from individual transformations or sectors will apply universally to other sectors or processes. There is still a shortage of convincing policies and practical strategies for the comprehensive management of transformations (meta-governance). Such policies and strategies would need to demonstrate how the numerous sub-processes, actions, tools and stakeholders can be adequately orchestrated, and what form of interplay between politics, civil society, business and academia might in practice be appropriate in the context of decision-making and further management of the transformation.

Below, possible approaches are proposed to structure the basic fields and options for action, and to explain them with reference to practical examples.

3.1. Transformation puzzle

In the context of the multi-level perspective (see Chapter 2), various patterns of transformation are discussed. An *intentional change in the landscape* is difficult to achieve and, as a rule, is only possible through international cooperation between several or many countries and actors, as with climate change. However, it may be assumed that through targeted observation, potential or emerging changes in the landscape can be identified and corresponding niche developments can be prepared or promoted when the time is right. This applies even to large-scale and sudden changes in the landscape. For example, it may be assumed, with some degree of certainty, that in the coming years, another serious reactor accident could occur, that refugee flows will continue to increase, that further major flood disasters can be expected, or that new financial and economic crises will necessitate the financing of major programmes at short notice.

A more promising approach than pursuing goal-oriented changes in the landscape is thus targeted management with the aim of encouraging social and technological innovations to emerge from their niches. Innovations can be developed outside the regime at first, and then consolidate with other innovations to form a new regime (see detailed discussion in Section 3.2.2.1).

There are *four key drivers* of transformations (WBGU 2011), whereby combinations of these drivers are required as a rule:

– **Vision**: e.g. sustainable development, energy system transformation, German reunification (see Section 3.2.3; on values and models, see Section 3.3.2)
– **Crisis**: e.g. the massive eruption of Mount Tambora in Indonesia in 1815, which caused worldwide harvest failures and triggered waves of emigration in subsequent years; numerous chemical problems and accidents in the 1960s to 1980s; threat to the ozone layer; Fukushima nuclear disaster
– **Knowledge**: Knowledge about demographic shifts, climate change (see Section 3.3.8)
– **Technology**: Numerous technological developments in the field of ICT (see Section 3.3.6.)

The three key drivers which can be *actively* influenced (vision, knowledge, technology) are discussed in subsequent chapters.

Section 3.2.4.1 considers whether it is possible to actively trigger tipping points in social systems.

A transformation can be managed in a goal-oriented and directional manner. However, it cannot be planned in detail; rather, it involves intensive processes of innovation, exploration and learning. Nonetheless, it must – as far as possible – be strategic and its individual steps must be planned, due allowances being made for a degree of uncertainty and scope for error. Unless an attempt at strategic planning is made, transformations to sustainability are unlikely to move in the desired direction, or will do so purely by chance.

The demands made of innovation management (see Section 3.2.2.1) demonstrate the need for conventional planning. Even open-ended exploration processes (Section 3.2.4) can and must be planned in terms of goal attainment, funding, timing, priorities and exit options. Ultimately, transformations are based on hundreds of individual innovations and actions, which – unlike the complex overall process of transformation – can and must be precisely planned. Examples are campaigns, product development, new business models, and the German Government's grid development planning in the electricity sector.

In view of the present state of theory and practice, describing the governance of transformations and outlining a strategic approach is difficult for two reasons. Firstly, very few publications set out practical steps to be taken in developing a transformation governance or transition management regime (see, for example, Rotmans & Loorbach 2009, WBGU 2011, Kristof 2010a and b). Secondly, there are very few examples of goal-oriented transformations to sustainability such as the energy transition (which is still under way).

There is, however, a wealth of practical experience and analyses of successful and failed major technological and social *innovations* from the past. Examples are the promotion of nuclear power since the 1960s; market liberalisation in the telecommunications, energy, transport and water sectors (since the 1990s); development of the Transrapid high-speed train system in Germany (1969 – 2009); the UN's sustainable development agenda since 1992 (e.g. Convention on Biological Diversity, UNFCCC, Millennium Development Goals); and the UN Decade of Education for Sustainable Development (2005 – 2014)[7]. The task becomes easier if planning focuses initially not on major transformation but on intermediate and smaller-scale transformations or portfolio innovations.

---

7 UN Decade with impact – 10 years of education for sustainable development in Germany. German Commission for UNESCO (DUK), Bonn 2014

Figure 3 Transformation puzzle

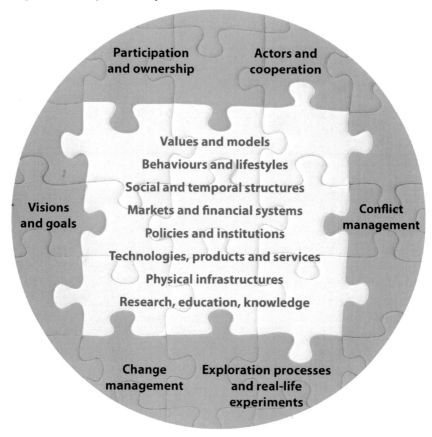

Source: Own depiction

The transformation puzzle shown in Figure 3 therefore provides guidance on

- the goal-oriented management of transformations to sustainability,
- the environmentally-oriented shift or enhancement of ongoing social innovations: the "greening of societal change" (Jacob et al. 2014).
- sustainability-oriented prevention of current or forthcoming *undesired* transformations,
- sustainability-oriented optimisation – and portfolio management – of innovations.

The term ***puzzle*** was chosen deliberately. There is a clear image and vision of what is to be achieved. The individual pieces of the puzzle are available and familiar, and there is an awareness that they have to combined with other pieces of the puzzle in a manner yet to be determined (exploration and learning processes) in order to make progress. The more clusters are formed from individual pieces, the clearer the picture. However, the puzzle is not a game but a challenge – and it is fraught with conflicts. Even if the venture is successful, the finished picture – transformation or individual innovations – will not be identical to the image in one's mind at the outset.

The eight inner pieces of the puzzle correspond to key fields of action/ social subsystems in which innovations and initiatives towards transformation progress in an interdependent or co-evolutionary manner. The six outer pieces of the puzzle represent the main process-related challenges facing the strategic actors involved in an intentional transformation. These process-related challenges may vary in the level of ambition that they require in the key fields of action, depending on the respective status quo.

The six process-related challenges are described below (Section 3.2). The eight subsystems are then discussed in Section 3.3.

3.2. Challenges arising in transformations

3.2.1. Actors and cooperation

It is still unclear – indeed, there is conflicting evidence – as to which actors actually drive an intentional transformation or, indeed, "proclaim" and manage it actively, as with the energy transition: The state? Civil society? Stakeholdercooperations?

Civil society is regarded as a driver: pioneers of change from within civil society are often involved at the start of a transformation. However, a key role is also assigned to the state. The "truth" probably lies somewhere in the middle and is likely to be phase-dependent. The stakeholder configurations are polycentric. Social and technological innovations by pioneers of change from civil society and companies are more likely to feature prominently at the start of a transformation, but as it evolves, state actors play an increasing role – by creating enabling spaces, providing financial support for innovations and stakeholder networks, adopting new laws and establishing new infrastructures.

With targeted transformations (which, like the energy transition and climate action, take place under pressure of time, unlike unintentional transformations), it is also doubtful whether and, indeed, how a *key actor* can be identified with the potential to initiate and facilitate this type of process in various subsystems, to outline targets and draft a strategy for implementation, to involve other actors and to advocate for measures in the various subsystems.

In principle, the key actor may be an influential social group (e.g. the trade unions, with their long campaign for a five-day week) or the state, because it can exert considerable influence on the various subsystems – through legislation, the provision of funding, taxation, and support for networking.

If the social innovation approach is pursued, however, it is clear that a large number of civil society, state and economic actors (but also stakeholders from academia and the arts) are involved in transformations, perhaps playing a variety of roles as the process evolves. Pioneers of change or other stakeholders interested in transformation will do well to keep an eye on "neighbouring" subsystems and process-elements in order to learn from developments there and generate synergies. What's more, various drivers and spoilers can be identified among the civil society and state actors from the outset. Some state actors may pursue similar goals as progressive civil society from an early stage. Examples are certain municipalities (e.g. the city of Freiburg; see Chapter 4), federal states (e.g. the State of Hesse in the context of chemical policy in the 1980s) and public authorities such as the German Federal Environment Agency – UBA (especially during the period 1976-1989 far from the Bonn seat of government and in the experimental political environment of the Berlin exclave).

Depending on the scale of the transformation, basic examples of state or political actors are a government, parliament, a federal state, study commission, ministry, or public authority (which then need to be equipped with a mandate for this area of work) and international agencies, e.g. UNEP.

Intentional transformation can, conceivably, also be supported by a central network or loose cooperation between (potential) change agents pursuing essentially the same goal, who consult each other, coordinate their activities or reference each other in a variety of ways. One example is the German chemical policy in the 1980s and 1990s (Held 1987), with actors Friends of the Earth Germany (BUND), the Oeko-Institut, Greenpeace, various citizens' action groups, the Federal Environment Agency, the State of Hesse, the Green Party, and some groups within the IG Metall trade union. Additionally there were books such as "Seveso ist überall" (Vahrenholt 1980) and "Chemie im Haushalt" (Öko-Institut et al. 1984), discussion platforms such as the Protestant Academies in Tutzing and Loccum, the Chemicals and Environment Information Service (ICU)[8] and the German Bundestag's Study Commission on Chemicals (formal title: Protection of Public Health and the Environment – Assessment Criteria and Perspectives on Sustainable Management of Material Streams in Industrial Society) (1992-1994)[9].

In conventional management processes in companies and policy-making, the identity of the manager(s) and key actors is clear. This is not the case during the initial stages of a transformation. What's more, these actors may change as the process continues, due to the long duration of the transformation. The relevant actors may be quite unaware of their role in shaping, influencing or facilitating the transformation, at least to begin with. Nonetheless, the first challenge in an intentional transformation is to ensure that (potential) actors are aware of their role and seek out like-minded people so that together, they can draft or at least coordinate their strategy. The puzzle approach supports this **polycentric management** of transformations through innovations and interventions and, in practice, enables individual actors to focus on separate pieces or clusters of the puzzle.

Figure 4 shows the typical actor groups. When selecting and approaching actors of relevance to the transformation or to individual innovations and interventions, it is important to be aware that these groups are by no means homogeneous.

---

8 ICU Informationsdienst Chemie & Umwelt, Journal published by the Bundesver-band Bürgerinitiativen Umweltschutz (BBU), Friends of the earth (BUND), Öko-Institut, Freiburg, 1984 - 1997,7/8

9 Study of the German Bundestag's Study Commission on Chemicals: „Schutz des Menschen und der Umwelt - Bewertungskriterien und Perspektiven für umweltverträgliche Stoffkreisläufe in der Industriegesellschaft"

Figure 4 Transformation actors

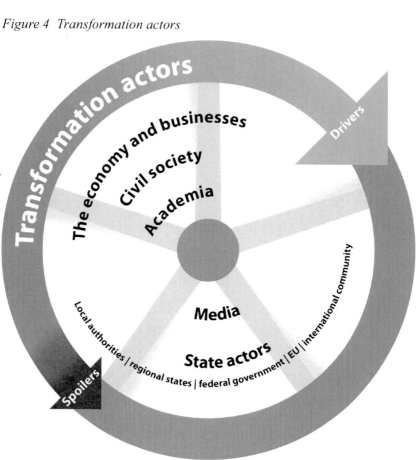

Source: Own depiction

In each of these groups of actors, there are drivers and spoilers, and their role can change – in either direction! – during the course of the transformation, in individual sectors in transition or at the various levels of governance.

The success of transformations depends on **cooperation among actors**; overall, participating actors need to contribute a good mix of organisational and individual skills: knowledge, power, resources, creativity, willingness to embrace innovation, dialogue skills, persuasiveness and process management skills. These pioneers of change are also known as early adopters, innovators, creative entrepreneurs and change agents. A wealth of information and recommendations are available on the various roles, selection and appropriate involvement of actors and inter alia the promoter model. The model, which is recommended by numerous authors (e.g. Kristof 2010a), aims to facilitate the interplay between various types of change promoter: technical promoters for knowledge-sharing; power promoters for overcoming unwillingness; and process and relationship promoters for communication and interaction. Besides there is a taxonomy of further references and recommendations, which is largely self-explanatory:

– niche actors; regime actors;
– local, regional, national, European and international actors;
– interpretative elite; facilitating elite; decision-making elite;
– market intermediaries; political intermediaries.

It is important to not only focus on niche actors but also on actors operating within the regime. They can play a dual role by supporting the existing system while also participating in niche activities. It may be particularly important to secure the backing of powerful regime actors in order to achieve broader legitimacy or mobilise support.

3.2.1.1. Case study: energy transition committees as an example of successful actor cooperation

The transformation discourse offers many general recommendations and typologies of actors and actor cooperation, but provides few examples of practical long-term cooperation. This section therefore briefly describes energy transition committees in Germany and their development as an example of successful actor cooperation, and identifies key success criteria (for a detailed discussion, see Grießhammer et al. 2015).

After the Chernobyl disaster, the Oeko-Institut – which had been working on the risks posed by nuclear power plants for some time – became a nationwide contact point for the media, concerned citizens and local authorities. In response to the flood of questions and "in order to counter the sense of isolation and paralysing fear and bring joint and massive pressure to bear on the relevant politicians and authorities ..." (Öko-Institut 1986), the Oeko-Institut called for the establishment of local, cross-party energy transition committees, referring, in this context, to its energy transition study published in 1980 (Krause et al. 1980). Within a few months, an energy transition network came into being, comprising around 400 committees set up by various types of organisation (peace, environmental and anti-nuclear initiatives, and concerned citizens' and parents' groups). The non-profit Oeko-Institut funded a part-time staff member to support the network and the individual committees over a 10-year period. Key elements of the work and activities undertaken by the energy transition network included:

- education, analysis and sharing of information, preparation of materials, lectures, seminars, trade fairs and conferences;
- supporting energy efficiency competitions and organising campaigns;
- hosting nationwide energy transition congresses (development of strategies, recommunalisation options, responses to the expiry of concession agreements, and actions on municipal and regional energy strategies and, later, climate change and renewable energies);
- municipal and regional projects on the ground (municipally-owned utilities; planning of local power generation plants; energy action plans; efficiency programmes; energy policy statements).

Ten years later (1996), 280 energy transition committees were still in existence. The number decreased significantly in subsequent years, partly because this area of work was becoming increasingly professionalised (founding of local energy agencies and businesses, shift towards larger NGOs, mainstreaming in politics and administration) and partly because the focus shifted to other activities (e.g. Local Agenda 21) and to climate action and renewable energies, with numerous associations and businesses being set up and support being provided for the construction of photovoltaic plants and wind farms. Nonetheless, some very active energy transition committees are still in existence, such as the *Arbeitskreis Klimaschutz und Energiewende Niedersachsen* (Lower Saxony Climate and Energy Transition Working Group), *Energiewende Saarland e.V.* (Saarland Energy Transition) and *Energiewende Rüsselsheim e.V.* (Rüsselsheim Energy Transition). The best-known example is probably the green energy supplier EWS – Elektrizitätswerke Schönau, which evolved from the *Eltern für atomfreie Zukunft* (Parents for a Nuclear-Free Future) energy transition initiative.

In the case of the energy transition committees, the key success factors were:

- the Oeko-Institut as an experienced organisation with a high level of commitment to transformation and credible objectives;

- the Oeko-Institut's publication of the book *Energiewende – Wachstum und Wohlstand ohne Erdöl und Uran* (1980), which set out the energy transition and provided a well-substantiated strategy as a basis for public debate and further development by municipal decision-makers (recommunalisation) and outlined new business models for innovative companies (least cost planning for municipal utilities);
- an environmental movement which was already actively engaged on energy and other issues, along with numerous creative and committed individuals and entrepreneurs;
- a window of opportunity as a result of the Chernobyl disaster, which had profoundly affected people across all sections of society on a very personal level;
- a clear strategy for cooperation and a well-established network;
- an (albeit modest) budget;
- targeted and cooperative PR.

To sum up: Congruent activities in several relevant subsystems (*values and models; behaviours and lifestyles; research, education, knowledge;* and, as the transformation progresses, *markets* and *policies and institutions*), a preparatory "blueprint" and its (substantive) development by various social actors underpin the success of cooperation in a transformation context.

3.2.2. Change management and innovation management

Prior to any consideration of management, the question which naturally first arises is who should be regarded as the manager(s) or active stakeholders (actors). This was discussed in detail in the previous sections. The following comments are relevant to all transformation-minded actors and, indeed, to some innovations, but they presuppose a higher degree of organisation and adequate capacities. Some recommendations relate primarily to state actors, especially those concerning financial support and legislation.

Change management comprises the coordination of processes, activities and actors, the identification of key leverage points, and the safeguarding of long-term and coherent support for transformation and a consistently enabling setting for sustainability. It also includes a suitable toolkit, e.g. internalisation of external costs, the promotion of pioneering activities and innovations, technical and social innovations, and real-life experiments/

living laboratories (for a description of methodology, see Wissenschaftsministerium 2013, p. 30). Linked with the core tasks of classic management (identification of goals and capacities; strategy development; organisation of implementation; monitoring of progress), change management must thus fulfil the following core requirements:

– definition of visions and goals
– regime analysis and identification of key fields of action
– scenario modelling and backcasting
– identification of key leverage points
– identification of (other) actors with similar goals for the purpose of cooperation
– targeted innovation management.

As the transformation progresses, further challenges arise, especially for state actors, in addition to *targeted innovation management* (Section 3.2.2.1): these challenges include ensuring citizen *participation and ownership* (Section 3.2.5), *organising exploration processes and real-life experiments* (Section 3.2.4), and managing conflicts appropriately (Section 3.2.6).

3.2.2.1. Targeted innovation management

In society's regime, many attempts are made to develop technological and social innovations. Many of them fail because the underlying concepts are too narrow in focus and the innovations themselves do not link in with developments within the individual subsystems (see also Bauknecht 2015).

This applies especially to technological innovations whose social embedding is neglected. Innovations generally progress in a systemic rather than linear manner, and their progression may be surprising and unplanned. Famous examples are computers ("I think there is a world market for maybe five computers." Thomas Watson, IBM Chairman, 1943[10]; "There is no reason for any individual to have a computer in his home." Ken Olsen, founder of Digital Equipment Corp., 1977[11]). And also the Internet, which was developed in 1962 and was originally intended to

---

10 https://de.wikipedia.org/wiki/Thomas_J._Watson
11 https://de.wikipedia.org/wiki/Ken_Olsen

be a communication tool for research institutes and later for the military, not a public network for the mass market. In some cases, innovations are successful and enjoy a surge in popularity, only to be replaced fairly quickly by others (fax machines are one example).

All successful innovations – whether individual innovations without systemic change, or portfolio innovations towards transformation – are embedded in societal structures or bring about change in these structures: "All innovation is social innovation" (Urry, 2011). For example, shared accommodation and lodging services such as Airbnb and new transport services such as rideshare schemes, Car2go and Uber, delivery services, etc. build on a new set of assumptions about the way people live, work and structure their time and on preconditions at other levels: Internet, smartphones, GPS, fast food, etc.

But how can innovations for transformation be identified and promoted? For **strategic niche management**, the following recommendations are made (and are particularly relevant to state actors):

– selection not only of technological innovations but also new social practices and arrangements and institutional rules (legislation, funding programmes);
– targeted selection and prioritisation based on systemic scenario and foresight processes, and the establishment of a National Office of Social Innovation, as in the US (UBA 2014);
– safeguarding and consolidation of innovations, e.g. adoption of protective measures;
– linkage and coordination of niche activities (nexus arrangements), e.g. through technical discourses, transdisciplinary research projects, cluster formation, interministerial working groups, Local Agenda processes or policy networks;
– parallel development of multiple and competing(!) niches, e.g. electric cars, hybrid cars, fuel cell vehicles, due to the open-endedness and uncertainty surrounding future developments.

Typical government funding mechanisms for innovations are R&D projects, legislation to promote innovation, e.g. the German Renewable Energy Sources Act, and specific cultural milieus which are receptive to new technologies or social practices and are willing to apply them at an early stage of development. Also the active setting of innovation targets, as with the Golden-Carrot-Initiative in the US, is possible. The initiative is based on government-sponsored competitions to foster companies' product innovations, based on environmental criteria, quality, cost and annual sales. The winning company receives substantial prize money amounting to around €10 million (Irrek et. al 2011).

When preparing innovations or support schemes, it is important to look beyond their specific development potential and analyse whether they:

- offer structure-changing solutions to problems within the regime,
- can be combined cumulatively with other innovations/subsystems (see Section 3.3) to form a mutually reinforcing innovation portfolio,
- are accepted by actors outside the innovation network (network embedding),
- offer general potential for upscaling and an increasing or incremental embedding in the regime.

Niche innovations may be particularly suitable in this context as they emerge, to some extent, independently of, and without selection pressure from, the regime (although they are, of course, influenced by it). *Innovations also emerge from within the* regime, but typically lack the potential to bring about structural change in the system. For example, in the prevailing energy regime, innovative boilers for coal-fired power plants have been developed (without exerting any pressure to change on the existing regime), whereas innovative wind energy technologies have emerged from the niche created by the Renewable Energy Sources Act (EEG).

Innovations can also emerge from *changes in the landscape*, e.g. a new financial crisis. In response to the global banking and financial crisis, economic stimulus programmes and stability packages amounting to around USD 2,000 billion were launched. In Germany, this amounted to USD 64 billion for two economic programmes (including the vehicle scrappage scheme[12]) and €400 billion for financial market stabilisation[13]. Such amounts would have offered enormous potential to directly foster new innovations.

Niche innovations can change if they become part of the existing or new regime (e.g. car sharing schemes are no longer organised as associations, as they were at the start, but as limited companies; renewable energies are increasingly required to provide system services as they feed

---

12 http://www.kfzabwrackpraemie.de/
13 German law to stabilize the financial market: http://dip21.bundestag.de/dip21/btd/ 17/083/1708343.pdf

more energy into the grid; flexible electricity loads are managed directly, as pioneer users cannot be relied on to provide management services on a broad scale).

The challenge, then, is to promote niches and innovations with the potential to support intentional transformation in particular. For example, the development of safe low-maintenance bicycles, e-bikes and charging stations could have been supported much earlier in order to encourage cycling rather than car use. The first fully functioning e-bike with a performance level comparable to that of modern e-bikes was registered in Germany in 1989!

Further development and upscaling of innovations, model projects and initiatives are extremely important. However, "projectitis" – support for a multitude of individual projects without any underlying strategy or coordination – is often a major weakness ("Projects never fail, projects never scale"). In their meta-analysis of 75 projects, Grießhammer et al. (2003) investigated the actions and measures typically adopted to promote sustainable consumption over a 10-year period (1992-2002). A key weakness identified in the initiatives – which were entirely successful if judged on their own merits – was the lack of an overarching strategy: sustainable consumption was not treated as a cross-cutting political task, and there was little networking between the initiatives. The majority of the proposals put forward in the study have not been implemented to this day.

3.2.3. Visions and goals

Visions are a key driver of transformations. The – generally indeterminate – start of a transformation is marked by criticism of the regime and, increasingly, ideas and concepts of more attractive alternatives, or visions. David and Leggewie (2015) draw attention, in this context, to the importance of participatory vision development, of communicating positive visions ("narratives") and of overcoming fears and resistance.

Ultimately, there is a shifting of values and models (see Section 3.3.2), with conflicts arising between the entrenched values and models that support the regime (e.g. car-friendly cities; the *Freie Fahrt für freie Bürger* anti-speed limit campaign) and new alternatives (e.g. the compact city; flexible mobility). The models and visions may be simple and clearly defined (e.g. the nuclear phase-out) or more complex (e.g. sustainable development, green economy, energy transition). The strong role played by visions and models is often underestimated. They describe a clearly more attractive alternative to the status quo and, for that reason, can often facilitate a new start and ease the way, which tends to be fraught with obstacles. They focus directly on citizens and voters and, with them, advocate for transformation. Through participatory vision development, visions and models are refined and improved in a collaborative process and can thus attract more supporters. If they remain diffuse (e.g. the green economy), they are less likely to inspire and motivate. The simpler the concept, the more convincing it is, and the easier it is to build on people's own experience. In the transformation literature, particular emphasis is placed on the "translation" of visions into "**narratives**"; in other words, into clear descriptions and images of a more attractive alternative. For this translation to be attractive, a matching language and matching image must be found (while being aware that classic marketing teaches us that there are no images that will suit everyone). Emotions determine communication with and the reachability of target group(s) in a pivotal manner.

The emergence or deepening of visions and societal goals can certainly be supported by the state. Examples are the international community's decisions on sustainable development and consumption, the German Chancellor's National Dialogue on Quality of Life[14] and the support from the German Environment Ministry (BMUB) and Environment Agency (UBA) for Sustainability 2.0 projects or resource-light lifestyles. However, for an intentional transformation to evolve from visions and narratives, they must be increasingly translated into clear goals in a strategic development process. In a transformation, these goals should be defined as precisely as possible. They must also be measurable so that progress can be evaluated.

If, during the transformation, quantitative targets are set by the state, they should be clear and uniform, with safeguards if possible to ensure that they remain in place across legislative terms and forthcoming elections. There are already two negative examples relating to the energy transition. The "historic" decision to phase out nuclear power (2002) was amended with the extension of nuclear plant lifetimes (2010), only to be swiftly renewed after Fukushima (2011). The decision to construct "electricity superhighways" was initially supported by Bavaria, but was rejected outright shortly before its local elections.

---

14 www.gut-leben-in-Deutschland.de

The setting of (transformation) goals is partly qualitative (e.g. healthy food, or reducing meat consumption) and partly detailed and quantitative (e.g. in the field of climate change and energy transition).

In a polycentric transformation, it is quite possible that different actors will pursue goals of varying levels of ambition. However, the goals should at least be broadly similar or complementary and form a "goal corridor".

Capacity issues (stakeholders' human capacities and time resources, and the budget(s) available to various actors) are rarely addressed but also need to be clarified.

3.2.4. Exploration processes and real-life experiments

Transformations can take decades. Goals can be set and pursued in order to guide the transformation in a desired direction. Under no circumstances, however, can the transformation be planned in detail. In particular, it is impossible to predict future technological and social developments with any certainty, and the longer the timeframe, the more this applies. For example, from today's perspective, it is impossible to say for sure how major electricity storage systems will develop in future, or which type of car will come to predominate in the next 20 or 30 years (which type of propulsion system; size; whether cars will be (conventionally) driver-driven, semi-autonomous or even remote-controlled). Transformations, therefore, necessarily involve exploration, learning and experimentation processes. Facilitating and shaping exploration processes means thinking about new ideas in new ways, guaranteeing a degree of flexibility, tolerating mistakes and learning from them, and keeping options open. According to Albert Einstein, ("Problems can never be solved with the same mindset that created them."), new mindsets are necessary. At the same time, in keeping options open, there is a trade-off between flexibility and secure conditions for investment (e.g. in the future "filling station network") or policy decisions (e.g. construction of electricity superhighways vs. the hope that large-scale and affordable electricity storage systems will be developed).

From a reflexive governance perspective (Voß et al. 2006), in strategies for managing transformations to sustainability, interaction spaces should *not* be confined to the conventional geographical, social or institutional parameters, but should be tailored to the specific problem. To that end, real-life (real-world) experiments/living laboratories are also proposed

(Schneidewind & Singer-Brodowski 2013)[15]. However, testing out new approaches must also mean being able to abandon ideas after trials have produced a negative result.

It has to be stressed that political experimentation needs courage on the part of decision-makers and confidence on the part of citizens – and vice versa!

The methodological concept of **living laboratories** has also been proposed for the transdisciplinary implementation of sustainability goals. The first living laboratories have already been established (see Section 3.3.8 for a detailed discussion). Also during product development real-world experiments have been conducted by companies for many years. For example, some detergent manufacturers trial new products in small countries, e.g. Liechtenstein and Luxembourg, before launching them in larger European or global markets. For the political arena, Bauknecht and Voß (in Praetorius et al. 2009) have proposed the concept of **regulatory innovation zones** as frameworks for innovation processes in grid regulation. As the concept is based on a statutory exemption, it must be well-justified and properly evaluated.

The Roadmap for the Smart Grids Platform for Baden-Württemberg[16] discusses the option of establishing regulatory innovation zones in order to test the suitability of various technologies and steering mechanisms. Participation would be voluntary and time-limited. These regulatory innovation zones must comply with the relevant legislation, e.g. the provisions of the German Energy Industry Act (*Energiewirtschaftsgesetz* – EnWG), and the principles of the EU internal market. The Federal Network Agency and state regulatory authorities are also to be involved in the development process.

Cities and regions in particular may be a suitable framework for innovations, for living laboratories and also for regulatory innovation zones. Here, there are many possible linkages, for example, to the transition town movement and sector-based initiatives, e.g. the development of renewable energy regions. In transport policy, based on the concept of regulatory

---

15 The "living lab" concept is somewhat problematic. Devised in the social sciences, it resonates with natural scientists and technologists as it is based on the familiar concept of a laboratory. However, it is open to misinterpretation by actors on the ground, who may think that they are "being used for experimental purposes".

16 http://um.baden-wuerttemberg.de/fileadmin/redaktion/m-um/intern/Dateien/Dokumente/2_Presse_und_Service/Publikationen/Energie/Smart_Grids_Roadmap.pdf

innovation zones, the broad-scale introduction of a 30 km speed limit in a municipality would be an example of a practicable and – in all likelihood – positive real-world experiment. A notable experiment in the transport sector was conducted in the South Korean city of Suwon, which has more than a million of residents. One district (4,300 residents and 1,500 cars) was encouraged to go car-free for a month. The scheme, which was supported by the mayor, was the result of consultations with the local community (Otto-Zimmermann & Park 2015)

3.2.4.1. Identifying tipping points in social systems

There is no doubt that major political events such as the German reunification in 1989 and the global financial crisis in 2007 play an important triggering or accelerating role for the system concerned. Such events – and new directional decisions based on them – are known as tipping points because the system changes fundamentally and irreversibly by "tipping" into a different state. The fascinating questions arise, whether in the course of a transformation such (societal) tipping points can be specifically identified and triggered in democratic systems in a democratic manner or whether it is at least possible to prepare for tipping points that are likely to occur (see also Brohmann & David 2015).

In simple terms, two types of tipping point can be identified in the context of transformations and sustainable change:

– *Unintended human-induced changes* in the Earth system: in the environmental sciences, the term "tipping point" is used to describe anthropogenic and now unstoppable damage to ecosystems and ecosystem services (see, in particular, Rockström et al. 2009) and the global climate system, posing major environmental and social risks and causing damage on a massive scale.
– *Intended changes* in social systems: the starting points for societal transition are analysed in various contexts in order to identify their dynamics of change (see, in particular, Wood & Doan 2003). The underlying idea is that in social systems, it is possible to actively trigger an irreversible shift into a desired new state.

The term "tipping point" has been in use at least since the late 1950s and is applied in various disciplines, e.g. epidemiology, clinical research, sociology and psychology. However, no comparative, systematic or empirical analysis of tipping point phenomena is currently available for any discipline; nor has an interdisciplinary analysis been undertaken. In a meta-analysis, the tipping point literature in nine different disciplines was studied and categorised. Due to the highly diverse use of tipping point concepts and hence the variety of interpretations of the thresholds at which systemic change is triggered, it was not possible to identify an ideal model. The definitions differ not only across disciplines but also within the various fields of social science research. As yet, no generally applicable criteria for identifying tipping points have been established in the scientific discourse. Therefore, there is still a lack of clarity on precisely what can be described as a tipping point that would, in functional terms, trigger societal change. What is clear, however, is that the subsystems or everyday areas with identified tipping point potential are highly context-specific.

It may be possible to work towards tipping points in subsystems, but due to time and causality problems, it is not possible to determine whether the intervention will take effect at the "right" time and whether the tools selected (e.g. incentive schemes, process support, regulations) will be successful ("governance into uncertainty"); in other words, whether they will indeed trigger the desired sustainable change. Research studies on the governance of social innovations are also very circumspect on this issue (Aderhold et al. 2015, UBA 2014). They underline the significance of the numerous (model-type) exploration and learning processes for the necessary development of new lifestyles and economic paradigms, but they cannot yet predict the success of any future scaling-up.

To sum up, the available concepts do not include any suggestions for the strategic use of tipping points and the associated policy instruments, which could be used to trigger the radical change processes that are desired, or to avoid undesired developments.

It is, however, possible to prepare for social tipping points that are anticipated or are highly likely to occur (e.g. the next financial or economic crisis) and develop a set of actions to support intentional transformation (e.g. proposals that are conducive to the energy transition and can be integrated into the next economic stimulus programme). These actions can be tested in laboratory situations and experimental contexts. From a research perspective, an *interdisciplinary methodology* for the analysis, evaluation and monitoring of tipping points for societal transformations is required.

3.2.5. Participation and ownership

Transformations require "a culture of attentiveness, a culture of participation, and a culture of obligation" (WBGU 2011). Firstly, a cultural shift in society, change processes and innovations towards sustainability and intergenerational justice should be supported. The aim here is to develop citizens' creativity, participation and ownership. Secondly, the involvement of many further citizens and organisations is extremely important for the success of a transformation. This is evident from the current example of the energy transition.

David and Leggewie (2015) have analysed how a society's cultural aspects must change in order to create potential for a transition to sustainability. Culture, in this context, is defined as "the normalities we grow up with, the patterns of thought and behaviour that have shaped us, the processes and routines that have become automatic to us" (Trattnigg & Haderlapp 2013, p. 113). So far, cultural shifts towards sustainability have mainly emerged at the local level. Society's fundamental attitude – in other words, the interest in and willingness to embrace changes in our practices, routines and the conditions that underlie them – can and should be promoted via broad-based social and political initiatives. The moon landing has already shown that it is possible to plan and implement "utopian" projects.

However, in order to embrace the new – and, in that sense, a sustainable future – citizens and stakeholders must break with the status quo, overcome path dependencies, and engage in new vertical and horizontal networking. This requires new forms of participation by civil society and new forms of cooperation with policy-makers, which must be negotiated. Citizens must take the initiative and explore and trial new practices. Policy-makers' responsibility is to create platforms of discourse on this process. Existing examples in Germany are the Citizens' Dialogue of the Environment Ministry, the Agenda-Setting of the Research Ministry, Citizen Science and the idea for a National Office of Innovation and Future Search. The municipalities in particular should provide arenas for cooperation and make serious efforts to involve citizens to a greater extent in future- and sustainability-oriented issues. Cultural change cannot and should not be imposed politically in a top-down manner, however. Rather, change is an iterative learning process which is initiated by communities themselves. A community which ventures to embark on a cultural transition to sustainability must therefore be clear in its own mind that its first task is to generate essential knowledge about change. Change agents must have the skills to link local debates with the global sustainability discourse, but above all, they need time and contacts if t.1ey are to progress the transition.

A further challenge is to involve many other citizens and the population as a whole, in addition to *directly active* citizens and stakeholders. This is important for two reasons. Firstly, citizens have many fears and concerns about future developments and major transition processes. Secondly, in transformation contexts, there may be many citizens who are affected directly or indirectly by changes in infrastructure, work processes, consumption, and the social environment. During infrastructure planning in particular, a reappraisal of existing participation processes is therefore required, with a focus on clearly defined goals, allocation of roles and responsibilities, a high level of inclusiveness, creative scope, an open-ended approach, clear alternatives, internal and external transparency, professional implementation of the process, and access to financial compensation.

3.2.6. Conflict management

Conflicts are to be expected in transformations. They can arise between regime actors and niche actors, among state actors, among niche actors, and among citizens and consumers. If the construction of new infrastructure is required, this is likely to create particular conflicts in densely populated, built-up countries such as Germany, as the old and new infrastructures will exist in parallel and "compete" with each other, at least for a transitional period. This can also create new "green-green" conflicts within the environmental movement. However, conflicts can also arise – as with Germany's energy transition – due to the division of responsibilities across various federal ministries and between the Federal Government and the German states, and due to the clearly divergent (political) interests between the federal and state levels and between the North German states (main producers of renewable-generated electricity) and the South German states (main electricity consumers).

In relation to conflicts, the transformation discourse provides various insights into their emergence, but offers little guidance on their practical management. Keywords are blockades and barriers, winners and losers, dealing with dissent, reaching compromises. Mainly, it is about distribution issues: who wins and who loses? The types of conflict vary too much to allow specific or practical recommendations on conflict management to be made. In general it is recommended to communicate the positive and new, prepare for potential conflicts, plan (!) for possible major conflicts, develop a conflict management strategy early on, target PR activities, mitigate impacts via transitional periods and/or compensation measures.

It is especially important, however, to win alliance partners by broadening the understanding of the problem (e.g. by framing a problem not only in environmental terms). Examples of the acquisition of further alliance partners are the Electricity Savings Check (*Stromsparcheck*) and Greenpeace's Foron campaign.

The Electricity Savings Check[17] is a federal scheme which provides advice on ways of saving electricity and heating to households receiving unemployment benefit under the Hartz IV legislation. Small energy-saving domestic appliances are also provided free of charge. This very successful scheme is implemented in cooperation with Caritas (social interest) and local authorities (financial interest, as the scheme reduces the utility-related benefits paid to Hartz IV households). Under the scheme, unemployed persons receiving Hartz IV benefits train as energy advisors (in both their own and the labour market policy interest).

Greenpeace's advance order campaign[18] was intended to encourage consumers to purchase the first CFC-free refrigerator, based on successful cooperation in 1993 between Greenpeace and Foron, a small manufacturing company in East Germany. The company had an economic interest, while many East and, indeed, West Germans supported the campaign for political and social reasons. As a result of the successful campaign, all the major Western German white goods manufacturers switched to CFC-free refrigeration technology in a matter of months.

Targeted negotiations are another option. An analysis of negotiated solutions as a cooperative approach to policy termination can be found in Heyen (2011) with reference to the decisions on the SPD-Green nuclear phase-out in 2001 and the ending of subsidies for German coal-mining. Both can be regarded as small-scale to intermediate transformations.

Even in a traditional management process within an organisation, internal and external **communication** plays an important role. Here,

---

17 http://www.stromspar-check.de/
18 https://www.greenpeace.de/themen/klimawandel/klimaschutz/der-greenfreeze-geschichte-eines-siegeszugs-0

however, the management can prepare its external communications and can, to a large extent, decide when it wishes to commence its public relations activities. In the case of an intentional transformation with the involvement of external actors, communication plays a key role in early participation and vision-building and then in the long-term transformation process. A dysfunctional public debate can substantially impact on a transformation process and even cause it to fail. In 2013 and 2014, for example, the impact of the EEG levy on domestic electricity prices was played up, and key additional information was not communicated (e.g. reduction in the average price of power on the electricity exchanges as a result of the feed-in from renewables; numerous exemptions for energy-intensive companies).

3.3. *Transformations in subsystems and key fields of action*

Transformations lead to structural changes in various subsystems (see Chapter 2). The precise definitions and demarcations of these subsystems vary in the literature. In essence, it is possible to identify eight subsystems and corresponding fields of action (see Figure 2) in which, in intentional transformations, goal-oriented, directional and mutually reinforcing initiatives and interventions should be implemented in a systems perspective:

- changes in the dominant values and models, such as the prevailing production and consumption culture, e.g. through transformation narratives, communication of positive visions, and alternative prosperity indicators;
- changes in individual behaviours and lifestyles, e.g. through information or pilot projects which aim to create positive models;
- changes in social and temporal structures, e.g. through new products or legislation (such as on working time, shopping opening hours);
- reform and reorganisation of unsustainable physical infrastrutcures, e.g. in electricity generation, the building stock;
- changes in markets and financial systems, e.g. through internalisation of external costs, new financing and business models;
- support for sustainable products and technologies by promoting innovations and niches, R&D projects and legislation (e.g. the Ecodesign Directive);

– in the realm of <u>research, education and knowledge</u>, support for trans-disciplinary transformation research in particular; expansion of education for sustainable development;
– use of new <u>policies and institutions</u> by the state to create an enabling environment.

The description of these eight subsystems[19] clearly illustrates the great extent to which behaviours and conditions, everyday culture, individual action and social structures, and the micro and macro level influence each other, and how important they are for the management of sustainable development processes and social innovations. Some subsystems are influenced more strongly by civil society (values and models, behaviours and lifestyles, social and temporal structures), while others are influenced to a greater extent by the economy (markets and financial systems, products, technologies and services) or the state (policies and institutions, infrastructures, and knowledge and education). However, parallel developments are always required in the other subsystems, along with cooperation between groups of actors, in order to facilitate society's transition to sustainability.

Systemic innovations can be unplanned and happen by chance, or they can be influenced in a targeted manner. Successes are most likely to be achieved if all the subsystems are addressed and interact. Changes in consumers' values, awareness and behaviour generally take place as systemic innovations in conjunction with other social developments.

A planned transition requires, on the one hand, a systems perspective which aims to transform social subsystems. On the other, it also requires an analytical approach which distinguishes between subsystems and fields of action in order to identify entry points for initiating change and interactions between the subsystems.

Looking at environmental protection and environmental and sustainable development policy over recent decades, it is clear that there have been many successful initiatives and actions at the individual levels. Some were coordinated on a multi-level basis, but the majority were stand-alone initiatives which did not lead to structural changes within society. It is important to note that parallel and "autonomous" social developments

---

19 For each of the subsystems, there are extensive specialist discourses that can not be presented here in depth. See on this Wolff et al. 2013, among others.

have occurred, which were not planned and, in all probability, could not have been triggered by individual actors.

3.3.1. Case study: the evolution of cycling – a partially planned systemic innovation

The evolution of cycling, as a significant contribution to daily mobility, illustrates the strong context-specificity of changes and the interplay between various levels, in terms of technology and social acceptance. The growth in the popularity of cycling has come about as a result of various mechanisms which have interacted and reinforced each other in time and space and across institutions. The following table illustrates, with reference to cycling, how a systemic innovation may evolve. It shows how the status of the bicycle/cycling has change over several decades: partly unplanned (driver: the keep fit movement) and partly driven by local environmental groups, NGOs and various government initiatives (see Table 1).

With hindsight, it is noticeable that despite the existence of initiatives and elements of planning to promote cycling, there was no systematic strategy for a transformation. It is clear how the developments interacted in the individual subsystems.

From a transformation perspective, the question which arises is how cycling would have developed if today's transformation-related knowledge had been available in the 1980s and if there had been much earlier and more intensive and targeted promotion of low-maintenance, safe cycles and e-bikes and expansion of the cycling infrastructure.

Below, the eight subsystems are presented individually and explained with reference to examples. Previous or potential initiatives and interventions that move in the direction of – or are counterproductive to – sustainability are identified.

Table 1 The evolution of cycling

Subsystems	Cycling – selected examples
Values and models	1980s: growing environmental awareness, leisure, keep fit movement; declining usefulness of cars in inner-city traffic; 1990s: surge in popularity of the Tour de France increases interest in cycling
Behaviours and lifestyles	Increase in cycling in the modal split from 9% (1976) to 15% (2011); households' bicycle ownership rises from 36.5 million (West Germany, 1980) to 69 million (Germany as a whole, 2009).
Social and temporal structures	Accepted average distances (time-dependent) are around 4.5 km at weekends and 3 km on weekdays. Changes might be possible due to rapid growth in the number of e-bikes.
Physical infrastructures	Considerable local and regional variations; high potential evident from cities known to be cycle-friendly, e.g. Münster and Freiburg: expansion of cycle paths and networks, fast cycle routes; cycles and local public transport have priority in traffic; 30 km zones in cities; cycles permitted to travel against the flow of traffic in some one-way streets; separate signage for cyclists; bicycle parking areas at stations and adequate cycle stands in city centres; cycle facilities on trains and local public transport.
Markets and financial systems	Increasing concentration of production; some cheap imports; falling costs of cycles in cost of living index.
Technologies, products and services	Increasing differentiation between bicycle types: city bikes, trekking bikes, sports bikes and, increasingly, e-bikes. Oeko-Institut initiative for safe, low-maintenance bicycles. Launch of bicycle rental schemes.
Research, educa- tion, knowledge	Overview of research findings: Ministry of Transport cycling portal; education materials
Policies and institutions	Promotion of infrastructure (see above); changes to commuter travel allowance (2001)

Source: Own compilation

3.3.2. Values and models

This subsystem consists of normative reference points such as values, social or statutory objectives, standards, attitudes, visions or ideas about what constitutes "the norm", and which apply either to society as a whole

45

or to individual areas of need. At the micro level, they are reference points for individuals; at meso level, they are shared reference points, e.g. for lifestyle groups; and at macro level, they provide a fundamental social and cultural frame of reference. Both the literature and practitioners often call for a shift in culture and mindsets towards sustainable lifestyles and consumption patterns. Changed or desired values are often translated into guiding visions. Representative public surveys are used in many cases to determine the level of general environmental awareness (BMBU/UBA 2015) or specific values and attitudes relating to the environment and how they may be changing. For example, in 2012-2014 the overwhelming majority of the population (90%, 90%, 89%) regarded public support for the energy transition as very important or important (BDEW 2014).

Visions are the shortest conceivable narratives. Visions may be aspirational (Martin Luther King: "I have a dream"), inspire courage (Kennedy: "Ich bin ein Berliner"), demand sacrifices (Churchill: "Blood, sweat and tears") or new policies (Brandt: "Make the sky blue again over the Ruhr", 1961), be a slogan for tens of thousands of protests ("Nuclear power? No thanks"), be associated with wellness (Slow Food; quality of life), be socially oriented (the trade unions' slogan "Saturday is Dad's family day"), or promote urban development or, indeed, urban decline over decades (car-friendly cities – the vision which guided urban planners in the 1960s) or encourage car use (the "Freie Fahrt für freie Bürger" anti-speed limit campaign).

In the 1970s and 1980s, a number of books on environmental topics had a considerable impact in Germany and acted as guiding visions (e.g. *Limits to Growth (Meadows et al. 1972), Energiewende, Chemie in Lebensmitteln (Katalyse-Umweltgruppe 1983)* and *Chemie im Haushalt (Öko-Institut et al. 1984)),* although in some cases, the vision was simply the negative alternative to the problem itself (no chemicals in foods …).

Films can also have a considerable influence on changing values (as for example "Die Wolke"[20]).

Serious visions which are difficult to translate, such as sustainable development and the 2° global warming target (generally set by political bodies), are less well-received by the public but can have a powerful impact in the political arena.

---

20 https://de.wikipedia.org/wiki/Die_Wolke_%28Film%29

In Germany's energy transition, four goals are pursued: the nuclear phase-out, the introduction of renewable energies, energy efficiency increases, and climate protection. For the first two goals, there are clear visions and visualisations. However, they are less clear for climate protection and very weak in relation to energy efficiency.

Examples of targeted interventions: the anti-nuclear protests (since the 1980s), the establishment of energy transition committees and their many years of work after Chernobyl (1986 ff.), the Rio Declaration (1992), the German Agriculture Ministry's campaign to promote the organic sector (2002), the requirement for German health insurance schemes to provide primary prevention services in order to improve health (Section 20 of Book V of the German Social Code)[21], the German Chancellor's National Dialogue on Quality of Life (2011), the transition town movement, and urban gardening (initiatives in recent years).

3.3.3. Behaviours and lifestyles

This category comprises (consumption) behaviour, daily routines and habits. The various levels can be distinguished, firstly, by the number of persons (individual action; shared group practices; societal practices) and, secondly, by one-off actions vs. lifestyle. The term "lifestyle" encompasses all (consumption) behaviour by an individual or group. People's actual behaviour and lifestyles often differ considerably from their values and awareness (e.g. environmental awareness vs. behaviour). A range of individual, psychological and infrastructural reasons are cited by way of explanation, but the main factors include unfavourable social and temporal structures, non-internalised environmental costs, and unfavourable tax conditions (e.g. conditions in other subsystems). This often leads to patch-work behaviour: citizens behave in an environmentally compatible manner in one sector (e.g. transport) but less so in another (e.g. food).

Various lifestyles are described in the Sinus-Milieus[22] (i.e. target groups that really exist – a model which groups people according to attitudes to life and ways of living) in consumer research, are advocated by environ-

---

21 https://www.gkv-spitzenverband.de/media/dokumente/presse/publikationen/Leit-
    faden_Praevention-2014_barrierefrei.pdf
22 http://www.sinus-institut.de/

mental organisations or are promoted through advertising (e.g. LOHAS – Lifestyles of Health and Sustainability).

Examples: Individual or average transport behaviour: car driver only, cyclist only, flexible use of modes, etc.

Examples of targeted interventions: Interventions, particularly in this subsystem, can vary considerably: flash mobs, urban gardening, free-ganism, information platforms and campaigns such as utopia[23] and EcoTopTen[24], eco-labels (e.g. Blue Angel) and sustainability labels. By promoting model projects, social marketing campaigns such as for organic labelling (2001), regulatory instruments such as speed limits, and by taxing consumption, the state can influence behaviour and lifestyles towards sustainability, thus creating enabling spaces for social innova-tions. However, the state often has an adverse effect (tax breaks for company cars, no speed limits, tax exemptions for aviation, etc.). It is noticeable, that the question whether the state should seek to influence behaviour and lifestyles is generally discussed in relation to environmental measures. A comparison: The one-off costs of the controversial but successful Biosiegel organic labelling campaign (see above) amounted to € 7.5 million (BMELV 2002)[25]. The tax breaks for aviation and VAT exemptions cost the German state around € 8,000 million (equivalent to more than 1000 Biosiegel campaigns) – year on year!

3.3.4. Social and temporal structures

Examples of structures which can act as social and cultural determinants include gender roles, living and working conditions, demographic changes, and social infrastructures such as kindergartens, school canteens and shopping opportunities. These are often linked to temporal structures, e.g. meal times, working hours, leisure time and vacations.

The transformation literature identifies a number of key time-related elements of transition processes, e.g. the duration of the transformation, windows of opportunity, and innovation diffusion processes. In relation to

---

23 http://www.utopia.de
24 www.ecotopten.de
25 http://www.pressrelations.de/new/standard/dereferrer.cfm?r=85132

social developments and natural processes, a range of dynamics, durations and temporal regimes can be observed:

- different **temporal regimes** (temporal logics, system times) in the various functional spaces (economy, politics, family, environment, society) and their evaluation;
- the effect of relatively narrow timeframes for government policy-making due to short legislative terms, which is diametrically opposed to long-term or, indeed, generational thinking (e.g. the vision of sustainable development);
- and thus the systematic consideration of different timeframes for measures to take effect (short-, medium-, long-, very long-term), as this entails "time-segmented" responsibility: processes whose impact will only be felt in the long term do not fit into legislative timeframes.

Windows of opportunity may be especially important in initiating or reinforcing transformations. Windows can be "opened" by major one-off events such as Fukushima. However, they can also arise in the problem stream (Kingdon 1995), e.g. due to new findings on climate change, or in the political stream (e.g. a change of government or a policy pursued for many years, such as Ostpolitik, 1968-1989). The global economic and financial crisis in 2007 can be seen as an unused window of opportunity because most of the resulting economic and support programmes were not environmentally oriented.

The appropriate speed of intentional transformations is a further topic. According to Kristof (2010a & b), this depends on the transition goal and on actors' conceptions of time. Excessive speed overtaxes most stake-holders' adaptive capacities. Excessively long processes, on the other hand, may well exhaust stakeholders and increase the likelihood that the project will be "talked out". The most beneficial approach is "rapid but not hasty implementation".

In intentional transformations, a dual challenge exists: defined goals should be achieved, but this must happen more rapidly than would be the case with an unintentional transformation. WBGU, for example, empha-sises the narrow window of opportunity and the great pressure to act in order to achieve the 2° climate target. Further time pressure arises because society's support for transformation can diminish over time due to other factors.

3.3.4.1. Case study: time affluence and the politics of time

The transition towards more sustainable lifestyles is already being lived and experienced in some niches within society. Broadly speaking, these niches are characterised by less polluting, more equitable employment, production and consumption patterns that create material and non-material prosperity. Time elements play a key role in new models of prosperity and the transformation to a culture of sustainability (Reisch & Bietz 2014).

In developed countries such as Germany, time aspects now play an important role in people's decisions on lifestyle, products and services; indeed, in some cases, they may weigh as heavily as the desired benefits or the cost of the product – and not only in more affluent circles. Lack of time, for all social groups, is one of the main reasons for suboptimal consumption decisions. The desire "to save time" and have control over one's time is almost universal. Many products and services promise to save the consumer time, at least compared with their competitors. At the same time, and paradoxically, the expansion of consumption increases the demands on our time. Furthermore, cost reductions in products and services are often associated with an increase in hidden time costs (beta version software; Internet travel booking systems; reduced service from telecoms providers, etc.).

For the public, it is not only about length or lack of time; it is also about time sovereignty and synchronicity. To promote sustainable lifestyles and implement the transformation, it is extremely important to understand consumers' and citizens' time-related needs. At the national and the international level (United Nations and OECD), prosperity, wellbeing and happiness have been topics of debate for some time, and the use and allocation of time are recurring issues in this context. In the German Government's National Dialogue on Quality of Life, time issues – such as the balance between, and value placed on, family, caring and working time – are of particular interest.

In consumer research, there is a long tradition of investigating the influence of time on consumption decisions. On the one hand, this is approached from the perspective of time perceptions and consumer benefit (= utility: process, goal and equipment utility) in relation to time. On the other hand, the significance of windows of opportunity – in other words, the right moment in time – in supporting sustainable consumption decisions or climate-friendly behaviour is also discussed. In Germany, the concept of time affluence was taken up in the 1990s, mainly in the context of research on new "post-material" lifestyles and "quality of life" (Reisch & Bietz 2014). Time affluence was, and is still, understood as the further development of conventional (material) prosperity and is mainly viewed in relation to sustainable consumption and production patterns. The concept is discussed in terms of the individual level (personal time affluence), the societal level (collective time affluence) and the institutional level, both as a means and an end, as a quantitative distribution problem and as a qualitative production problem.

Research on the ecology of time identifies key dimensions of time affluence as follows: the chronometric dimension (duration), the chronological dimension (time situation), time sovereignty or time autonomy, the right or opportune moment (kairos), synchronicity, appropriate speed, rhythm, and consideration of one's own and the system's time (Reisch & Bietz 2014). Ultimately, it is about time autonomy and time sovereignty: having sufficient time per time-use to fulfil obligations and desires (available time) and, overall, for a work-life balance between one's own needs and the demands made on one's time.

Reisch and Bietz (2014) discuss in detail the significance of time aspects in policy-making and expressly call for a "politics of time". And indeed, policy-makers often make decisions which have time implications, thereby greatly intervening in societal and individual time arrangements. One example can be found in the energy transition. For many years, the energy supply system relied primarily on base load power plants (nuclear and lignite) and surplus electricity, especially at night. Electricity was therefore much cheaper at night, so industry switched, as far as possible, to night-time tariffs and electric heating was promoted. In the power supply system now being established, in which there is abundant or, indeed, an oversupply of solar power at midday and in summer, a paradigm shift is taking place.

Examples of social and temporal structures: working time, meal times, responsibilities, gender roles, access to local public transport, access to telecoms/broadband networks, kindergarten/creche opening hours, shop opening hours, warranty periods.

<u>Examples of targeted interventions</u>: bank holidays, adoption of summer time, the switch from nine-year to eight-year grammar school[26], legitimate claim for a place in a crèche for all children younger than three years[27], mail server switch-off by companies at weekends (no reading of work-related emails during leisure time), time-dependent tariffs, labelling of appliances' lifetime electricity costs.

3.3.5. Markets and financial systems

This subsystem comprises market structures (e.g. degree of concentration, globalisation) and market processes such as supply, demand and prices of goods and services, but also market regulation (e.g. anti-trust, liability and competition law), the international financial system and general financing issues. Distinctions can be made between local, regional, national and global markets, each of which have their own specific features.

Markets are constituted through autonomous market processes and through interventions (or omissions) by the state.

The funding of new infrastructures (see Section 3.3.7), financial support for new product R&D (see Section 3.3.6) and cleanup or dismantling processes can reach a critical mass in intentional transformations and lead to falling levels of approval on the part of the public and industry. The problem is built in, in that a transformation which would otherwise take place over a much longer period of time is accelerated in a goal-oriented manner and, as a consequence, investments are made over a much shorter timeframe. In the debate about the energy transition, it is therefore proposed that the costs be extended over time via a fund model.

<u>Examples</u>: financial transactions, business models, internal company requirements (e.g. return on investment – ROI), high-speed stock trading, organisational forms such as cooperatives, etc.

New business models and organisational forms have triggered a shift in ownership structures in the energy sector. Within a few years, the structures shifted from four energy supply companies with a few hundred

---

26 Graduation after 12 years of school is called G8
27 http://www.familien-wegweiser.de/wegweiser/stichwortverze-
    ichnis,did=38652.html

power plants (1990s) to around 1.3 million power-generating facilities, most of them owned by private citizens, funds, farmers, municipal utilities, SMEs etc., in 2014. The four major energy suppliers accounted for only a 5% renewable energy share in 2013 (Agentur für Erneuerbare Energien 2013).

Examples of targeted interventions: subsidies or reduction of subsidies, internalisation of external costs, reporting obligations, tariff structures, outsourcing of informal activities, mandatory labelling of (high) product operating costs.

3.3.6. Technologies, products and services

This subsystem comprises individual products <u>and</u> services, as well as generic technologies. If they offer attractive benefits (= utility), they can be a key driver of transformations. Products and technologies generally develop as a result of research by pioneers and companies, and their further development is supported – or sometimes curbed – by private or state actors. There has been little analysis, to date, of <u>desirable</u> technologies, products and services in intentional transformations.

A notable example of the successful promotion of technology is the support provided for photovoltaics and wind power under the German Renewable Energy Sources Act (EEG). In a newspaper campaign in 1993, the energy industry claimed: "... renewable energies such as solar, hydro and wind will be unable to meet more than 4% of our energy needs, even over the long term" (SZ 1993). In 2014 the share of renewables was already up to 27% and the once high generation costs of photovoltaics were reduced by almost 90% in just 15 years. In sun-rich countries, photovoltaic technology is now the most affordable source of electricity.

Examples: computers, mobile phones, eBay, Facebook, genetic engineering, photovoltaics, wind energy, carsharing, e-bikes.

Examples of interventions: government-funded R&D programmes, start-up support for energy-efficient products, the Blue Angel eco-label[28], the

---

28 https://www.blauer-engel.de/de

Ecodesign Directive[29], the Golden Carrot programme[30], financing programmes for building retrofits provided by the state-owned KfW bank[31], economic stimulus programmes.

3.3.7. Physical infrastructure

Physical infrastructures comprise relatively durable material structures which influence or even dominate the spaces in which actors operate.

Examples: the road and rail networks, filling station infrastructure, telecoms/broadband network, large linked industrial complexes, the electricity grid, the buildings stock, expansion of the rapid charging network.

Examples of targeted interventions: promotion of renewable energies (German Renewable Energies Act), power grid expansion, KfW loans for energy efficiency improvements to buildings, expansion of cycling infrastructure, expansion of telecoms/broadband network.

3.3.8. Research, education, knowledge

This subsystem comprises science and research, including its institutional arrangements, relevant education programmes at various levels, and, more generally, the knowledge needed to support transformations. Here, a distinction can be made between problem-related and system-related knowledge (understanding of the current situation and problems), knowledge of reference points and targets (i.e. knowledge of target situations that may offer solutions to the problem) and knowledge of transformations or actions (knowledge of methods to achieve the goal). At micro level, this means the knowledge and level of education of individuals or groups; at macro level, it means social organisation of knowledge generation and sharing via the research and education system.

---

29 http://www.evpg.bam.de/de/richtlinie/index.htm
30 http://www.nrel.gov/docs/legosti/old/7281.pdf
31 http://www.bundesregierung.de/Webs/Breg/DE/Themen/Energiewende/Energies-paren/CO2-Gebaeudesanierung/_node.html

The body of knowledge about problems, perspectives and solutions can be a key driver of transformations.

At the forefront of knowledge and research is the **promotion of transition/transformation research and transformative and transdisciplinary science**. The social-ecological research conducted in the BMBF program having the same name gave essential impulses (Jahn et al. 2000, BMBF 2007).

WBGU (2011) analysed research requirements and knowledge generation relating to transformations and distinguishes between transformation research and transformative research. These are not fully formed and clearly defined branches of research with unified theories and methodologies, but terms used to denote research activities in various disciplines and research contexts – activities which will need to be expanded and focused in future. Transformation research aims to obtain information about the progression of, and factors influencing, societal transition (conditions, drivers, obstacles). Transformative research, by contrast, deals primarily with problem-solving strategies by means of which processes can be actively advanced and managed. The transition between the two branches of research is fluid: knowledge from transformation research can and should be utilised for transformative purposes. In both branches of research, transdisciplinarity is a key (although not the only) method (briefly defined as interdisciplinary study of practical issues together with practitioners).

Transdisciplinarity and transformation research already feature in the research strategies and programmes launched by the German Research Ministry and the EU (Horizon 2020[32]). The need and opportunity to develop and trial socio-technical innovations in real-life contexts are reflected in the methodological concept of living labs (Schneidewind & Singer-Brodowski 2013), mainly a subject of discussion in Germany and recently adopted by the government of Baden-Württemberg. Living labs can be understood as the targeted management of niches in which socio-technical innovations can gain a foothold with a degree of protection, but also under real-world conditions. Living laboratories are, above all, an innovative methodological concept for science, situated between purely participatory observation and controlled laboratory conditions. Living laboratories are intended to offer scientists better access to the active

---

32 http://ec.europa.eu/programmes/horizon2020/

management of social transformation processes and pursue a transdisci-plinary approach, as various scientific disciplines and practical knowledge converge in the design and operation of the living lab.

In their article *Politikrelevante Nachhaltigkeitsforschung* [= Policy-relevant Sustainability Research], Jahn and Keil (2012) have summarised the resulting requirements that must be met by researchers, funding practice and funding providers, and their findings apply to a large extent to research on transformations towards sustainability.

The scientific organisations, for the most part, are not well-placed to meet the new challenges. There is also considerable professional and insti-tutional scepticism in the traditional academic community towards trans-formative science. In a recent statement, the *German Council of Science and Humanities* underlines the shared responsibility of the scientific community and science policy-makers in identifying and overcoming the "major challenges facing society" and makes various recommendations for action (Wissenschaftsrat 2015).

Proceeding from the debate on **citizen science**[33] in the English-speaking world, more intensive citizen participation in science and research is now being discussed at EU level as well. Here, a distinction can be made between strong citizen science, in which interested citizens and representative organisations participate in the formulation of research questions, are involved in living labs, and contribute to the development of research programmes (co-production, co-design), and weak citizen science, in which citizens collect data for research purposes, e.g. weather, bird or butterfly data, or data on the state of the world's coral reefs[34] (Finke 2014, Dickinson & Bonney 2012).

Examples of targeted interventions: science policy and research funding (see above), full-day schools, curricula, the UN Decade of Education for Sustainable Development, adult education programmes.

3.3.9. Policies and institutions

Policies and institutions comprise, firstly, practical steering mechanisms such as imperatives and prohibitions, financial incentives, and information

---

33 http://buergerschaffenwissen.de/citizen-science/wie-funktioniert-citizen-science
34 http://www.reefcheck.de/

tools for the regulation of collective interests, and, secondly, institutional and organisational frameworks (government bodies, competences, separation of powers, democratic processes, legislation). All the other subsystems are influenced via regulations, subsidies and steering mechanisms. Measures can take place at various political levels (e.g. local authority, federal state, national or international level; see Section 3.3.9.1 below). However, they should, as far as possible, be coordinated, coherent and mainstreamed across departments.

Beyond individual policy interventions, more fundamental changes to political frameworks are possible, e.g. through amendments to the Basic Law (e.g. inclusion of environmental protection as a government objective), trade regimes (WTO, planned TTIP) or transfer of competences between institutions/levels. In line with this new thinking, Reisch and Bietz (2014) explore the importance of time aspects in policy-making and expressly call for a "politics of time". And indeed, policy-makers make many decisions about time, thereby massively intervening in societal and individual time arrangements.

Examples of targeted interventions were discussed in the previous chapters. The state can influence the other subsystems and change individual subsystems in full or in part. However, to achieve genuine success, it requires the support of societal actors: citizens and voters, businesses and social groups.

3.3.9.1. Multi-level governance

Beside the multi-level perspective used to describe the general progression of transformations outlined in Chapter 2, various analyses and theories have long existed in political science on multi-level governance, i.e. the interplay between various political levels – local, regional, state, EU and UN level – and the interactions between these levels and non-state actors (detailed discussion see Bauknecht 2015).

In the German **energy transition**, for example, there are many participating institutions and actors: at the "classic" political levels, they include the EU, the Federal Government, the German states, the counties (e.g. designation of priority areas for wind power plants), many small and large companies, funds and cooperatives (with a total of 1.3 million power-generating facilities; see Section 3.3.5), environmental and consumer organisations, business associations, trade unions, the crafts and trades, citizens' groups, etc., including new configurations (such as a coalition of local authorities and citizens' initiatives opposed to the planned electricity superhighway in Bavaria).

The various political levels have different tasks and policy-making/ governance options, both in mainstream politics and in the context of transformations. The interaction is not necessarily hierarchical, and some levels may be omitted (e.g. in global networks of cities for climate action, the nation-states are "left out"). The linkage with social transformation processes has rarely been analysed, with the exception of the prominent role played by local authorities.

Local authorities are regarded as particularly suitable places and promoters of niche developments for transformations; the city of Freiburg is a good example (see Chapter 4 for a more detailed discussion). In upscaling innovations and progressing transformations, however, the over-arching political levels, with legislation and financial support, are vital. The further development of the multi-level governance approach to trans-formations is important as the coordination of diffuse political control, also across state actors, is a major challenge in transition contexts.

Within the political sciences, there are various interpretations/concepts of political governance: the traditional narrow understanding of "govern-ment" focuses on state actors and steering from the top down. The modern, broader understanding of "governance" includes private and economic actors and networks, as well as formal and informal regulatory structures, and is thus ideally suited for developing options for action in transformations. On the other hand, of course, it also harbours the risk that initiatives will be blocked or watered down. Cooperative development of problem-solving strategies with or via the objects of governance (e.g. NGOs or other non-state actors) should lead to more efficient and appro-priate policy programmes and outcomes at the various levels, as the right actors can come together in a targeted manner to deal with the problem and address the issue at the relevant level.

4. Green City Freiburg

The need for a sustainable and equitable transformation to sustainability is becoming increasingly urgent. Local authorities in particular are assigned an important role in this context and are regarded as key actors in promoting niche innovations. Initiatives that are trialled at first at the local level can, if they are successful, later be scaled up to larger entities and to the national and international level.

With reference to the city of Freiburg, the powerful influence that environmental groups and social movements can exert at local level, their networking and the importance of niche developments for later upscaling are described below (upscaling refers here to individual innovations, such as energy-plus buildings, not upscaling of Freiburg as a green city).

Due to specific economic, geographic, political and regional factors, Freiburg and its surrounding area emerged early on as Germany's "environmental capital", and Freiburg is regarded by the international community as a model city. Every year, more than 25,000 professionals from around 45 countries visit Freiburg to learn about its environmental initiatives and best practice projects (Green City Freiburg 2015).

Freiburg has won numerous awards and has featured prominently in the rankings for being a particularly environmentally friendly and sustainable city: German Environmental Capital (1992), European Public Transport Award (1995), recognition as a Region of the Future (2000), Dubai International Award (UN-Habitat) for Freiburg's Vauban district (2002), European OSMOS Award (2007), European Green Capital Award (2009), European City of the Year, Urbanism Area (2010), Sustania 100: City of the UN Decade (2011) and the German Sustainability Award (category: cities) (2012). Numerous sustainability-related innovations, including legislation, have been developed in Freiburg, often serving as models of best practice and exerting considerable influence (Grießhammer & Hilbert 2015).

4.1. Pioneers of change

Almost all actions were initiated by individual pioneers, local environmental groups and coalitions, and later also by research institutes and
companies (backed by visionary individuals), with growing support from
local parties, civil society and the municipal administration. The local
setting offered opportunities for constructive exchange – and fruitful
competition – and created a climate conducive to social and environmental
innovations which became established and attracted further pioneers.
There was informal cooperation via personal contacts and structured cooperation through organised networks. In 1992, FAUST e.V. (Arbeitsgemein-
schaft Freiburger Umweltinstitute = Association of Environmental Institutes in Freiburg) was set up. The founder members included the Oeko-
Institut, BUND – Friends of the Earth Baden-Württemberg, the
Ökomedia-Institut (organiser of the first environmental film festivals),
ICLEI – Local Governments for Sustainability, the Bundesverband
Bürgerinitiativen Umweltschutz (BBU) (= Federal Association of Environmental Citizen Initiatives) and its Water Working Group, Freiburg
University's Institute of Environmental Medicine and Hospital Hygiene,
which is renowned throughout Germany, and other smaller and equally
committed environmental institutes, e.g. Forschungs- & Beratungsinstitut
Gefahrstoffe GmbH (FoBiG), Hydrotox and the Freiburg Institute for
Environmental Chemistry. In 2003, the Association was integrated into a
larger organisation, Ecotrinova. In parallel, many private citizens were
actively engaged, including the renowned solar architect Rolf Disch, and
Georg Salvamoser, founder of Germany's first production plant for solar
modules. It was from Freiburg that the campaign to establish energy transition committees, of which there were ultimately 400 across Germany,
was launched, which did much to initiate energy system transformation
(see Section 3.2.1.1).

In Table 2 (below), the left-hand and centre columns summarise the
model projects and innovations which took place in Freiburg and assign
them to the eight subsystems (see Section 3.2). The right-hand column
lists the corresponding interventions by the state. Of course, other than in
individual cases (e.g. Electricity Savings Check pilot project and funding
programme), it is impossible to make a direct link between Freiburg-based
or local pilot projects and subsequent government measures. For legislation to be adopted, the projects must be replicated in other municipalities

and at other levels, accompanied by political pressure and appropriate policy changes as a result of elections.

The table shows the key initiatives in the field of energy and transport, but there have been many initiatives in other sectors, e.g. in protection from harmful chemicals, or in organic farming, commerce and catering.

Table 2 2 Sustainability innovations in Green City Freiburg

Subsystems	Innovations and initiatives in/from Freiburg and region	Subsequent action by the state
Values and models	Since the 1970s: anti-nuclear protests in Freiburg and region (plans to build Wyhl nuclear power plant); municipal council votes in favour of phase-out 1980: publication of the Oeko-Institut's book on energy transition (*Energie-Wende*) 1984 Bestseller: *Der Öko-Knigge* and 1984 Bestseller: *Chemie im Haushalt* 1986 Ban on legendary uphill motor racing; replaced with car-free Sunday and later by a cycle fun run 2002 Freiburg becomes the first German city to elect a Green mayor	**2011 Second government decision to phase out nuclear power in Germany** **2011 Adoption of the Federal Government's Energy Concept to promote renewables**
Behaviours and lifestyles	Steady increase in the share and absolute number of cyclists in city traffic 2008 Electricity Savings Check for low-income households – pilot project 1999-2007 Citizen contracting – a student-parent-teacher consortium funds the energy upgrading of a comprehensive school (Staudinger Gesamt-Schule): model for many other similar projects across Germany From 2003 to date: consumer platform EcoTopTen[35] lists energy-efficient appliances and compares their overall costs	**From 2009 to date: Energy Savings Check funding programme** **2005 EU Ecodesign Directive**

---

35 http://www.ecotopten.de

61

Subsystems	Innovations and initiatives in/from Freiburg and region	Subsequent action by the state
Social and temporal structures	Since 1993: Vauban – a sustainable model district that adheres to environmental and social standards[36]; model for many other urban districts across Germany	
Markets and financial systems	1987 Founding of the first major mail order company specialising in green products (Waschbär) Since the mid 1980s: many new solar and wind power systems funded by Freiburg citizens 1996 Founding and construction of Germany's first solar module production plant Founding of the international marketing company SAG 1996 Low-energy light bulb campaign (Meister Lampe): municipal utility company distributes low-energy bulbs free of charge to its customers, recouping the costs via electricity prices (Roos et al. 1996)	**1999-2003: 100,000 Roofs Programme: promotion of photovoltaics From 2000 Renewable Energy Sources Act and amendments**
Technologies, products and services	From 1981 Further development of photovoltaics Founding of the Fraunhofer Institute for Solar Energy Systems 1989 Registration of Germany's first fully developed e-bike From 2000 Development of energy-plus buildings, Germany's first energy-plus district; first apartment block to be renovated to passive house standard	**Statutory measures –see above Funding programmes for building energy efficiency improvement, e-mobility, etc.**

---

36 http://www.vauban.de/

Subsystems	Innovations and initiatives in/from Freiburg and region	Subsequent action by the state
Physical infrastructures	Strong expansion of tram and regional rail network; linkage with Germany's first environmental ticket (affordable, large catchment area, covers all forms of public transport) Expansion of cycling infrastructure (cycle paths and parking areas, preferential rules on local roads, fast cycle routes) Traffic calming in new residential areas	
Research, education, knowledge	1975 Founding of Volkshochschule Wyhler Wald (adult education centre) on the site for the planned nuclear power plant, which is occupied by protesters 1977 Founding of the Oeko-Institut (non-profit; no start-up funding from government; transdisciplinary focus) 1984 Founding of the Ökomedia-Institut; hosts annual international film festivals 1986 Founding of the Ökostation Environmental Education Centre From 2013 Umweltkonvent – annual environmental convention, with 70 – 80 international environmental laureates	**From 2013 German and European funding programmes (BMBF, EU's Horizon 2020) focus on transdisciplinary projects Baden-Württemberg's Science Ministry launches competition for first living labs**
Policies and institutions	At local level, inevitably limited. Examples are the land-use plans for the new districts (Vauban and Rieselfeld). 1986 Freiburg is one of the first German cities to set up an environmental office	

Source: Grießhammer & Hilbert 2015

Through further action at local level (e.g. transport, photovoltaics, residential development, and construction projects), individual innovations have been expanded and professionalised, and have thus becomes models for adoption and upscaling at national and international level. Looking back, the importance of these first initiatives, the active commitment of individuals, and the diffusion through stakeholder cooperation become clear. It is also apparent that it can take some time for innovations to become established at national and international level, but with patience and persistence, success can be achieved. It is also evident that the resonance and long-term impact of such initiatives are generally much greater than the initiatives themselves realise at the time.

4.2. *Further need for action in Freiburg*

Notwithstanding all the enthusiasm about the many successful initiatives and model projects, it is important to note that environmental protection in Freiburg still offers considerable scope for improvement. Despite numerous energy initiatives, energy consumption is still high, as are greenhouse gas emissions and traffic congestion (e.g. due to a large number of commuters from the wider region); there is also considerable pressure to build housing, and a high level of land take. Although Freiburg and its residents are undoubtedly much more environmentally aware than most other cities and the average citizen in Germany, per capita consumption and emissions here are still far too high compared to other countries. According to a study by the Oeko-Institut, Freiburg will have to reduce its energy consumption by around 50% and massively increase the use of renewable energies in order to be carbon-neutral by 2050. There is still a lot to do, and there is still plenty of scope for new pioneers of change – in Freiburg and elsewhere.

5. Policy recommendations and further research

The following proposals are of particular relevance to policy-makers and public authorities, but they can – and should – also be adopted and utilised productively by civil society, business and researchers.

Strategic development

- Classification and prioritisation of current, desired and undesired transformation processes
- Targeted activities to destabilise undesired transformations (e.g. abolition of environmentally harmful subsidies and elements of the TTIP that pose a risk to the environment and democracy)
- Greening of current transformations that were not induced by environmental policy (greening societal change)
- Support for cultural change; mobilising support for societal transition towards sustainable consumption by promoting cultural change and increased participation.
- New funding models for transformations (e.g. to promote renewables within the EEG framework)
- Preparation of measures (blueprints) that promote transformations, with a focus on possible/probable windows of opportunity (e.g. financial crisis/economic stimulus programmes)
- Stronger political focus on exploration and learning processes, facilitation of regulatory innovation zones, real-life/real-world experiments and living labs
- Stronger focus on time aspects of transformations: ending short-termism in society and politics, e.g. through family-friendly working time and opening hours, longer-term policy- and decision-making across legislative periods.
- If resistance arises, a positive approach should be adopted, based on the assumption that change is needed, expressed through rejection of outdated patterns and structures. Resistance can also be a key indicator that the wrong pathway has been adopted and must be corrected.

Opposition to the loss of existing privileges can be mitigated by compensation measures or through negotiation.

Institutional support

– Establishment of a National Institution of Social Innovation and regular Future Search events (UBA 2014), together with targeted niche and innovation management for goal-oriented transformations
– Strengthening of international cooperation on transformations wherever these are internationally configured and supported (e.g. climate protection is an EU-wide goal, but the nuclear phase-out is not)
– Civil society groups and organisations have, in recent years, made a substantial transformative contribution in the field of sustainable development (e.g. collaborative consumption). In order to achieve even greater success, these groups and organisations should focus more strongly on underpinning their strategies and on networking and should receive more support for new approaches
– Stronger and timely corporate involvement in transformations; support for new business models; especially for SMEs; time-limited criteria-based subsidies with clear exit options; competitive tendering for product innovations (model: Golden Carrot initiative in the US[37]).

Science policy and research needs

– Science policy: greater focus on transformation research and transdisciplinary sustainability research (see recommendations made by WBGU and German Council of Science and Humanities on research policy and bodies); funding of living labs
– The great complexity of transformative research, especially in view of the numerous practical demands arising in the individual fields of research, necessitates the further development of transdisciplinary methodologies, together with quality assurance and research experience that transcends individual models. This includes, in particular:

---

37 http://www.nrel.gov/docs/legosti/old/7281.pdf

- Theoretical elucidation of interactions among subsystems
- *Mechanisms to support stakeholder cooperation* (who initiates it, how can transformation-minded actors come together, and how can viable agreements be achieved?)
- *Development of interdisciplinary methodologies* for analysis, evaluation and monitoring (e.g. tipping processes)
- *Methods to research* the possible embedding of new time structures (e.g. in political and corporate decision-making)
- *Proposals on coordinating diffuse political responsibilities* across all policy areas (including beyond state actors)
- *A toolkit for long-term political agreements* (how can they be safeguarded across legislative terms and changes of government?)
- *Methods to incorporate exploration processes* and regulatory innovation zones into politics and legislation
- Comprehensive empirical research that evaluates and compiles the findings and impacts of past transformation processes – specific to individual fields of action, wherever possible
- *New financing arrangements* for transformations.

In the following publications, further recommendations can be found: Bauknecht 2015; Brohmann & David 2015; David & Leggewie 2015; Reisch & Bietz 2014; Heyen 2013.

6. Looking ahead

The list of research needs and the policy recommendations show that in addition to the further governance of the energy transition, there is much to do in the years ahead: refining transformative methodologies, establishing frameworks for transformation processes (governance) and, not least, defining or influencing the thematic and strategic direction of intentional and non-intentional transformations. This is already apparent in the current industrial revolution, which is harmlessly described as the "Internet of things" but will lead to massive economic and social upheavals. As yet, there is very little sign of a social agenda in this technology-driven transformation, and it is quite unclear how the few goals which have been defined – such as data protection and data security – can be safeguarded in the planned fully networked world.

In parallel with the energy transition in Germany, intentional transformations are taking place and are at various stages of development. Examples are:

Sustainable development – in reality, the ultimate "Great Transformation" – proclaimed at the United Nations Conference on Environment and Development (UNCED) in Rio de Janeiro in 1992 and reaffirmed in the 2030 Agenda for Sustainable Development in late 2015. It is progressing, but very slowly – hardly surprising in view of the massive challenges and the need for agreement among some 200 countries. It remains a difficult message to convey.

There are many signs that a food transformation (also encompassing agriculture and livestock management) has already begun, and various initiatives have been launched. This is likely to be "officially" proclaimed as an intentional transformation in the next few years.

The same applies to a transport transformation. This will need to go far beyond a top-down shift to electromobility. Indeed, in decades to come, it is likely to evolve in directions we can hardly imagine today. Google's self-driving car and the package drone may be portents of what lies ahead.

We hope that this book will help the many committed individuals and organisations working in the environment and sustainability sector to gain a better understanding of current and future transformation processes, and that it will provide inspiration for their own strategies and actions. We're with Bertolt Brecht: "Be realistic: demand the impossible." We hope you will join us!

References

Aderhold et al. 2015: Aderhold, J.; Mann, C.; Rückert-John, J.; Schäfer, M.: Experimentierraum Stadt: Good Governance für soziale Innovationen auf dem Weg zur Nachhaltigkeitstransformation. Zentrum Technik und Gesellschaft TU Berlin und Institut für Sozialinnovation e.V., UBA-Texte 04/2015

Agentur für Erneuerbare Energien 2013: Agentur für Erneuerbare Energien, 2013. Trend Research - Stand 04/13. Available online at: www.unendlich-viel-energie.de

Bauknecht 2015: Bauknecht, D.: Gesellschaftlicher Wandel als Mehrebenenansatz, with contributions from Matthew Bach., UBA-Texte 66/2015

BDEW 2014: Energie-Info. BDEW-Energiemonitor: Das Meinungsbild der Bevölkerung (2012, 2013, 2014)

Bechmann 1987: Bechmann, A.: Landbau – Wende. Gesunde Landwirtschaft. Gesunde Ernährung. Fischer-Verlag Frankfurt 1990

BMBF 2007: BMBF – Bundesministerium für Bildung und Forschung: Sozialökologische Forschung. Rahmenkonzept 2007

BMELV 2002: BMELV – Bundesministerium für Verbraucherschutz, Ernährung und Landwirtschaft, Presseerklärung vom 15.01.2002

BMBU/UBA 2015: Bundesministerium für Umwelt, Naturschutz, Bau und Reaktorsicherheit, Umweltbundesamt: Umweltbewusstsein in Deutschland – Ergebnisse einer repräsentativen Bevölkerungsumfrage, Berlin 2015

Brohmann 1996: Brohmann, B. (Hrsg.): 10 Jahre nach Tschernobyl. Projekte für eine andere Energiepolitik, Öko-Institut Freiburg 1996

Brohmann & David 2015: Brohmann, B.; David, M.: Tipping-Points, Öko-Institut and KWI, UBA-Texte 67/2015

David & Leggewie 2015: David, M.; Leggewie, C.: Kultureller Wandel in Richtung gesellschaftliche Nachhaltigkeit, Arbeitspapier KWI, Essen 2015

Dickinson & Bonney 2012: Dickinson, J.L., Rick Bonney, R.: Citizen Science: Public Participation in Environmental Research. Cornell University Press, 2012

Eberle et al. 2006: Eberle, U.; Hayn D.; Rehaag, R.; Simhäuser, U.: Ernährungswende. Eine Herausforderung für Politik, Unternehmen und Gesellschaft. oekom verlag, Munich 2006.

Eppler 1975: Eppler, E.: Ende oder Wende. Von der Machbarkeit des Notwendigen, Kohlhammer-Verlag, Stuttgart 1975

Finke 2014: Finke, P.: Citizen Science: Das unterschätzte Wissen der Laien. oekom verlag, Munich 2014

Geels 2002: Geels, F.W.: Technological transitions as evolutionary reconfiguration processes : a multi-level perspective and a case-study. Research Policy, 31(8-9), pp. 1257–1274, 2002.

Green City Freiburg 2015: Green City Freiburg. Wege zur Nachhaltigkeit. Published by Freiburg Wirtschaft Touristik und Messe GmbH & Co. KG. Freiburg, 2015. Available online at http://www.freiburg.de/pb/site/Freiburg/get/640887/GC-Brosch %C3%BCre_D-2014.pdf

Grießhammer 1992: Grießhammer, R.: Szenarien einer Chemiewende, Freiburg 1992

Grießhammer et al. 2003: Grießhammer, R.; Brohmann, B.; Ebinger, F.; Gensch, C.-O,; Henseling, C.; Quack, D.; Michelsen, G.; Godemann, J.: Erfassung, Analyse und Auswertung der Aktionen und Maßnahmen zur Förderung des nachhaltigen Konsums durch gesellschaftliche Akteure im Rahmen der nationalen Verständigung. Öko-Institut and Lüneburg University, Freiburg/Lüneburg 2003

Grießhammer et al. 2015: Grießhammer, R.; Brohmann, B.; Hilbert, I.: Energiewende-Komitees als Beispiel für erfolgreiche Akteurskooperationen einer Transformation, in preparation

Grießhammer & Hilbert 2015: Grießhammer, R.; Hilbert, I.: Green City Freiburg, in preparation

Hesse 1995: Hesse, M.: Verkehrswende: Ökologisch-ökonomische Perspektiven für Stadt und Region, Metropolis-Verlag Marburg, 1995

Heyen 2011: Heyen, D.A.: Policy Termination durch Aushandlung: Eine Analyse der Ausstiegsregelungen zu Kernenergie und Kohlesubventionen. In: Der moderne Staat 4:1, pp. 149-166, 2011

Heyen 2013: Heyen, D.A.: Auswertung des aktuellen Forschungs- und Wissenstandes zu Transformationsprozessen und -strategien, with contributions from Dierk Bauknecht, Working Paper, Öko-Institut, Freiburg/Darmstadt/Berlin, 2013

IPCC 2007: IPCC – Intergovernmental Panel on Climate Change: Fourth Assessment Report – Climate Change, 2007

Irrek et al. 2011: Irrek, W.; Seifried, D.; Grießhammer, R.: Finanzielle Unterstützung der Produktentwicklung und Vermarktung hocheffizienter, energieverbrauchender Produkte. Öko-Institut in Kooperation mit Institut für Energiesysteme und Energiewirtschaft der Hochschule Ruhr West & Büro Ö-Quadrat Freiburg, 2011

Jacob et al. 2014: Jacob, K.; Graaf, L.; Bär, H.: Transformative Environmental Policies. Discussion Paper (Draft version of 23 June 2014) for 11th ESDN Workshop Berlin, 26 June 2014

Jahn et al. 2000: Jahn, T.; Sons, E.; Stieß, I.: Konzeptionelles Fokussieren und partizipatives Vernetzen von Wissen. Bericht zur Genese des Förderschwerpunkts Sozial-ökologische Forschung des BMBF. Studientext Nr. 8. ISOE. Frankfurt

Jahn & Keil 2012: Jahn, T.; Keil, F.: Politikrelevante Nachhaltigkeitsforschung. Frankfurt/Berlin, 2012

Katalyse-Umweltgruppe 1983: Katalyse-Umweltgruppe Köln (ed.), Chemie in Lebensmitteln, 9. Auflage, 1983

Kemp & Loorbach 2006: Kemp, R.; Loorbach, D.A.: Transition management: a reflexive governance approach, in Voß, J.-P.; Bauknecht, D., Kemp, R.: Reflexive governance for sustainable development, 2006

Kingdon 1995: Kingdon, J.W.: Agendas, Alternatives, and Public Policies. Longman (2nd edition), 1995

Krause et al. 1980: Krause, F.; Bossel, H.; Müller-Reißmann, K.-F.: Energie-Wende. Wachstum und Wohlstand ohne Erdöl und Uran. Ein Alternativ-Bericht des Öko-Instituts Freiburg. Fischer-Verlag 1980

Kristof 2010a: Kristof, K.: Models of Change: Einführung und Verbreitung sozialer Innovationen und gesellschaftlicher Veränderungen in trans-disziplinärer Perspektive. vdf Hochschulverlag 2010

Kristof 2010b: Kristof, K.: Wege zum Wandel: Wie wir gesellschaftliche Veränderungen erfolgreicher gestalten können. oekom verlag, Munich 2010

Loorbach 2007: Loorbach, D. A., Transition management: new mode of governance for sustainable development (pp. 1–328). Erasmus University Rotterdam, 2007

Meadows et al. 1972: Meadows, D.; Meadows, D.H.; Zahn, A.; Milling, P.: Die Grenzen des Wachstums. Bericht des Club of Rome zur Lage der Menschheit, 1972

Öko-Institut et al. 1984: Öko-Institut, Katalyse, VUA, BUND (Hrsg.): Chemie im Haushalt, Hamburg 1984

Öko-Institut 1986: Aufruf zur Gründung von Energiewende-Komitees, In: Brohmann, B. (Hrsg.) 1996: 10 Jahre nach Tschernobyl. Projekte für eine andere Energiepolitik, Öko-Institut Freiburg S. 18

Otto-Zimmermann & Park 2015: Otto-Zimmermann, K.; Park, Y.: Neighborhood in Motion – one neighborhood, one month, no cars, Jovis Verlag 2015

Praetorius et al. 2009: Praetorius, B.; Bauknecht, D.; Cames, M.; Fischer, C.; Pehnt, M.; Schumacher, K.; Voß, J.-P.: Innovation for Sustainable Electricity Systems – Exploring the Dynamics of Energy Transitions (Series: Su.). Physica Verlag Heidelberg 2009

Reisch & Bietz 2014: Reisch, L.; Bietz, S.: Zeit für Nachhaltigkeit – Zeiten der Transformation: Elemente einer Zeitpolitik für die gesellschaftliche Transformation zu nachhaltigeren Lebensstilen, UBA-Texte 68/2014

Rockström et al. 2009: Rockström, J. et al.: Planetary Boundaries: Exploring the Safe Operating Space for Humanity. In: Ecology and Society, 14(2), pp. 1-33, 2009

Roos et al. 1996: Roos, W.; Schüle, R.; Seifried, D.: Evaluierung der stromwirtschaftlichen Auswirkungen des Energiedienstleistungs-Programms der FEW für die Haushaltskunden 1996 (Meister Lampe), Freiburg 1996

Rotmans & Loorbach 2009: Rotmans, J.; Loorbach, D.A.: Complexity and Transition Management. In: Journal of Industrial Ecology, 13(2), pp. 184–196, 2009

Schneidewind & Singer-Brodowski 2013: Schneidewind, U.; Singer-Brodowski, M.: Transformative Wissenschaft: Klimawandel im deutschen Wissenschafts- und Hochschulsystem. Metropolis Verlag 2013

Smith & Raven 2012: Smith, A.; Raven, R.: What is protective space? Reconsidering niches in transitions to sustainability. Research Policy, 41(6), 1025–1036. 2012

Steffen et al. 2015: Steffen, W.; Richardson, K.; Rockström, J.; Cornell, S.E.; Fetzer, I.; Bennett, E.M.; Biggs, R.; Carpenter, S.R.; de Vries, W.; de Wit, C.A.; Folke, C.; Gerten, D.; Heinke, J.; Mace, G.M.; Persson, L.M.; Ramanathan, V.; Reyers, B.; Sörlin, S. (2015): Planetary boundaries: Guiding human development on a changing planet. In: Science 347(6223): 1-41. 2015

SZ 1993: Süddeutsche Zeitung Nr. 152, 1993. Advertisement by the utility companies Badenwerk Karlsruhe, Bayernwerk München, EVS Stuttgart, Isar-Amperwerke München, Neckarwerke Esslingen, PreussenElektra Hannover, RWE Energie Essen, TWS Stuttgart and VEW Dortmund

Trattnigg & Haderlapp 2013: Trattnigg, R., Haderlapp, T.: Zukunftsfähigkeit ist eine Frage der Kultur – Hemmnisse, Widersprüche und Gelingensfaktoren des kulturellen Wandels, oekom verlag 2013

UBA 2014: Umweltbundesamt (Ed.). Soziale Innovationen im Aufwind – Ein Leitfaden zur Förderung sozialer Innovationen im Aufwind, Berlin 2014

Urry 2011: Urry, J.: Climate Change and Society, Polity, Cambridge, UK, 2011

Vahrenholt 1980: Vahrenholt, F.: Seveso ist überall: die tödlichen Risiken der Chemie, Frankfurt am Main 1980

Voß et al. 2006: J.-P. Voß, D. Bauknecht; R. Kemp (Hrsg.): Reflexive governance for sustainable development. Edward Elgar Publishing 2006

WBGU 2011: German Advisory Council on Global Change (Wissenschaftlicher Beirat der Bundesregierung Globale Umweltveränderungen, WBGU): World in Transition – A Social Contract for Sustainability. WBGU Flagship Report 2011

Weizsäcker et al. 2009: v. Weizsäcker, E.U.; Hargroves, K.; Smith, M: Factor Five: Transforming the Global Economy through 80% Improvements in Resource Productivity. Taylor & Francis, 2009

Wissenschaftsminist. 2014: Ministerium für Wissenschaft, Forschung und Kunst Baden-Württemberg (Ed.): Wissenschaft für Nachhaltigkeit – Herausforderungen und Chancen für das baden-württembergische Wissen-schaftssystem, Stuttgart 2014

Wissenschaftsrat 2015: Wissenschaftsrat, Zum Wissenschaftspolitischen Diskurs über große gesellschaftliche Herausforderungen. Drs. 4594-15, Stuttgart 2015

Wolff et al. 2013: Wolff, F.; Barth, R.; Brunn, C.; Fischer, C.; Grießhammer, R.; Heyen, D.: Suffizienz im Alltagsleben – Konzept, Bedarf, Potenziale und politische Steuerungsmöglichkeiten, Berlin/Freiburg/Darmstadt 2013

Wood & Doan 2003: Wood, B.D.; Doan, A.: The Politics of Problem Definition: Applying and Testing Threshold Models, in: American Journal of Political Science, Vol. 47, No. 4 (Oct. 2003), pp. 640-653.